Published by
PEACOCK BOOKS
7/22, Ansari Road, Darya Ganj,
New Delhi-110002
Phones : 2327 3880, 23275880, 23280451
Fax : +91-11-23285873
Web : www.atlanticbooks.com
E-mail : info@atlanticbooks.com

Branch Office
5, Nallathambi Street, Wallajah Road,
Chennai-600 002
Phones : +91-44-64611085, 32413319
E-mail : chennai@atlanticbooks.com

*Peacock Books is an imprint of
Atlantic Publishers and Distributors (P) Ltd.*

ISBN: 978-81-248-0188-8

Printed in India at Nice Printing Press, A-33/3A, Site-IV,
Industrial Area, Sahibabad, Ghaziabad, U.P.

Dedicated to

The victims of the Gujarat Earthquake of 26 January 2001,
The heroic people of Kutchch, the Armed Forces,
NGOs, industrial houses, and all others
Who worked selflessly to provide rescue,
Relief and rehabilitation.

PREFACE

The earthquake, which hit Kutchch and other parts of Gujarat in the morning of 26 January 2001, sent out a grim reminder that in spite of all our technological advancements, we are powerless against the wrath of Nature. Our helplessness is increased manifold due to our ignorance about the extent of our vulnerability and our inability to take basic precautionary measures, which can reduce the extent of damage. The situation is further aggravated by the apathy of the Government and the bureaucracy to develop an effective disaster mitigation and management policy and plans. Thus, thousands of lives were lost because the rural houses were not earthquake resistant. Over Rs. 100 crores worth of ammunition was destroyed in the ammunition depot at Bharatpur because a few lakhs were not available for grass cutting. Over 3.5 crore people remained cut off from the outside world during disasters because we could not spend money to establish foolproof communication systems. Hundreds of people died because they could not be rescued in time.

India is a disaster-prone country. Every year, lakhs of people are rendered homeless because of earthquakes, floods and cyclones. Property, produce and productivity worth hundreds of thousands of crores of rupees are lost. We are happy to rebuild but are unwilling to take effective measures to prevent damage. Much of Holland is below sea level, but no one hears of major floods there. There have been over a dozen major earthquakes in our country since independence. But fortunately or unfortunately, most of these largely affected rural areas and the city-bred elite never felt threatened enough to take serious notice of the danger. To them—the film stars, the politicians, the rich and the famous and the bureaucrats—disasters are occasions when one can collect and donate funds, give profound interviews, gain publicity and get rid of old clothes. Within a couple of months, the disaster and its victims are largely forgotten. A few government servants, consultants, contractors and NGOs struggle to complete the rehabilitation work.

The vulnerability of towns, as demonstrated by the damage that occurred at Ahmedabad, Surat and Bhuj, has given a wake-up call. An earthquake of magnitude seven and above anywhere near Delhi or Mumbai could leave over 20 lakhs killed. For years, our permissive society has been tolerating corruption. The rich and the powerful are above law. Building laws are flouted. Illegal constructions have sprung up all over the place. Who knows when we will have to pay the price!

The aim of this book is to increase the awareness of people about earthquakes, the risks involved and what can be done to mitigate the effects of such a disaster. It also explains the methods of inspection and repair of earthquake affected buildings. The book is not a theoretical treatise on earthquakes. It is meant to serve as a useful guide for the readers to know about the causes and consequences of earthquakes and how they can protect themselves and their property from these disasters.

Col. Bhaskar Sarkar, VSM, Retd.

CONTENTS

PART I

Understanding Earthquakes

1

PAST EARTHQUAKES AND SEISMIC MAPPING OF INDIA

Information about earthquakes or tremors in the Indian subcontinent of more than 5 on the Richter Scale since about year 1800 are available with the Meteorological Department of India and the Indian Society for Earthquake Technology. The details of less than magnitude 5 are generally not available because adequately sensitive equipments to measure lesser tremors are not available in adequate numbers with the Meteorological department. Would you like hazard a guess as to how many tremors of magnitude 5 and above have been felt on in India these 170 years from 1800 to 1971. The number is about 210 (the list is attached at the end of this chapter) or an average of approximately 1.2 per year. These exclude foreshocks and after shocks and earthquakes in the Andaman and Nicobar Islands. Is it not strange that in spite of such a high frequency of earthquakes in the country, the public awareness about earthquakes is almost nil. This is perhaps because most of the earthquakes have been in the remote rural regions of the country and the rich and the powerful have not been affected. Before the television came, news about small earthquakes never caught the public attention.

Devastating Earthquakes of the Past

India has seen a number of devastating earthquakes in the past. Brief descriptions of these earthquakes are given in the succeeding paragraphs.

The Rann of Kutch Earthquake of June 1819

This earthquake occurred on 16th June, 1819 at about 6.45 PM. The epicentre was near Bhuj and its magnitude was about 7.5 on the Richter scale. The earthquake left over 1,500 people dead and damaged properties from Ahmedabad in the east to Porbandar in the west and Jaisalmer in the north. The towns of Bhuj and Anjar suffered heavy damage. Over 7,000 houses were destroyed in Bhuj alone. In Anjar, the houses built on low rocky ridges suffered little damage. The houses on the slopes leading to the springs and swamps were severely damaged. During the first three months, not a single day passed without aftershocks. There after the frequency of aftershocks decreased gradually. The aftershocks reportedly continued for eight years. Another remarkable phenomena associated with the earthquake was the visible faulting on the earth's surface which extended for 80 Kms. An earthen bund called "Allahbund" having a maximum height of 6.15 m and length of 80 Kms was formed. The area to the south of the bund subsided and formed a depression 55 Kms long and occupying an area of 514 sq. Kms. This depression was filled by sea water and became a lake.

The Assam Earthquake of 1897

The Assam Earthquake of 12th June, 1897 was one of the largest in the world. The earthquake occurred at about 5.15 PM local time and was responsible for over 1,500 deaths. The intensity was estimated to be more than 8.7 on the Richter scale. It led to almost total destruction of all brick and stone buildings in all the major towns in the area including Shillong, Goalpara, Guwahati, Dhubri and Tura. Landslides occurred on a large scale in Meghalaya and many fissures developed. At a number of places, the beds of streams tilted, resulting in rivers changing their courses. Even the bed of the Bramhaputra river was affected and caused large scale flooding in the later part of the year. There were a large number of aftershocks whose epicenter was apparently spread over a large area. Two very strong aftershocks were felt the next day. The intensity of these shocks declined but continued for at least 10 years.

The Kangra Earthquake of 1905

This earthquake took place at about 4.20 AM on 4th April, 1905. The epicenter was near Kangra. Its magnitude exceeded 8 on the Richter scale. The earthquake killed over 20,000 people. It was felt over an area of 4,16,000 sq. Km. No visible faults were produced.

Nepal – Bihar Earthquake of 1934

The epicentre of this earthquake lay below the alluvium slightly to the north of Darbhanga in Bihar. There were three zones of maximum intensity. The largest was an elongated tract of approximately 128 Kms from Motihari to Madhubani. The second was at Monghyr on the south bank of the Ganges and the third was near Kathmandu in Nepal. The towns of Monghyr, Motihari, Madhubani and Bettiah were almost completely destroyed. A belt of slumping extended from Bettiah in the north-west to Purnea in the south-east took place. Most buildings in the area of subsidence tilted and were damaged. There were many fissures through which large quantities of water and sand were thrown up thus ruining standing crops and making areas unfit for cultivation. There were extensive landslides in the Himalayan region. More than 10,000 people were killed. The earthquake was felt over an area of 48,60,000 sq. Kms.

The Assam Earthquake of 15th August, 1950

The Assam Earthquake of 1950 had its epicentre in Arunachal Pradesh near the junction of the borders of India, Myanmar and Tibet. The earthquake measured over 8.5 on the Richter scale. Long fissures were produced. The level of river Lohit and all other streams rose. Extensive landslides took place. Though the epicentre lay in a largely uninhabited area, the earthquake caused great destruction to the region and particularly to the sub-divisions of North Lakhimpur, Dibrugarh, Jorhat, Sibsagar. Arunachal Pradesh suffered heavy damage. Over 532 persons were killed. Rail and road communications were totally disrupted due to subsidence and large fissures. The bed of the river Bramhaputra rose and caused extensive flooding. The shocks were felt over an area of 29,00,000 sq. Kms. There were a large number of aftershocks. The largest one measured 7 on the Richter scale.

Anjar Earthquake of 1956

The Anjar earthquake took place on 21st July, 1956 and measured 7 on Richter scale. It devastated the town and adjoining areas.

Koyna Earthquake of December 1967

The earthquake took place at about 10.51 PM on 10th December, 1967 and measured about 6.5 on the Richter scale. The epicentre was located about 3 Kms south of the Koyna Dam at southern Maharashtra in the midst of the Western Ghats. A large number of houses in Koynanagar, a small township near the dam, were destroyed and 177 people lost their lives. The main damage was confined to a belt of about 8 Kms but the earthquake was felt as far away as Surat to the North and Hyderabad to the east. The earthquake was preceded by foreshocks, the first of which occurred on 4th June, 1965. The aftershocks continued for about 3 years. The magnitude of the earthquake and the damaged caused was not exceptional. However, the earthquake drew attention of many experts for two reasons. The first was that the area had no past history of earthquakes and was not considered earthquake prone. The second was that some experts opined that the earthquake was due to construction of the Koyna Dam and impounding water in the reservoir. However, this view was not shared by the team of scientists from UNESCO, who studied the earthquake.

Tehri Earthquake of 1971

This earthquake in the area surrounding the Tehri Dam caused extensive damage to Tehri town and villages in the region. Over thousand people were killed and thousands were injured. Thousands of houses collapsed or were severely damaged.

Uttarkashi Earthquake of 1991

This earthquake of magnitude 6.6 on the Richter scale struck in the early hours of October 20, 1991. The epicentre was about 30 Kms from Uttarkashi in Uttaranchal. Over 1,000 people were killed and thousands were injured. Over 30,000 houses collapsed or were severely damaged. Uttarkashi, Chamoli, Pauri, Tehri and Dehradun districts were affected. Maximum damage was at Uttarkashi where over 700 people were killed.

Latur Earthquake of 1993

An earthquake of about 6.3 on the Richter scale struck Marathwada region of Maharashtra in the early morning hours of 30th September, 1993. The epicentre of the earthquake was near the village of Killari, about 50 Kms south of Latur. The earthquake killed over 8,000 people, left over 20,000 injured and destroyed about 2,00,000 houses. The area within 40 Kms radius of Killari suffered extensive damage. The shocks were felt at Pune and Mumbai, which are over 400 Kms away. Marathwada region had no history of earthquakes and was not considered earthquake prone. However, the earthquake was preceded by foreshocks for over two years. The aftershocks continued for over three years. The massive loss of life and property was not due to the intensity of the earthquake but due to the traditional type of houses constructed in the area with mud, stones and heavy mud roofs, which were extremely vulnerable to earthquakes. The foreshocks, which preceded the earthquake, were not taken seriously and the people were not advised on the weakness of their houses in case of earthquake.

Vulnerability of States

Let us try to compare the state wise occurrence of earthquakes from the year 1800 to approximately 1971, the data for which is readily available. In addition to the numbers given below, there have been 32 earthquakes in Nepal, 96 in Tibet, 10 in Bhutan, 49 in Myanmar and 3 in Bangladesh. The state wise data up to 1971 and including a few prominent subsequent ones are given in the table on next page.

Andhra Pradesh is not very susceptible to earthquakes. There have been only two. The first was between 5 to 6 on the Richter Scale on 12 Oct, 1959 with epicentre near Guntur. The second was on 13th and 14th April, 1969 with its epicentre near Warangal and measured 6 to 6.5 on the Richter scale.

Arunachal Pradesh is the most earthquake sensitive of the Indian States. About 26 earthquakes had their epicentre in this small state. It has also been shaken by many earthquakes with their epicentre in Bhutan, Tibet and Burma. The first major earthquake was in 1906 and had its epicentre in the hills along its border with Burma. It measured 7 on the Richter scale and damaged some areas of Assam up to Sibsagar. This was followed by another major one on 12 Dec., 1908 with its epicentre about

No.	Name of State	Earthquake 5-5.9	Earthquake 6-6.9	Earthquake 7-7.9	Earthquake 8 and above	Total
1.	Andhra	1	1	–	–	2
2.	Arunachal	9	11	5	1	26
3.	Assam	19	4	3	–	26
4.	Bihar	3	–	–	–	3
5.	Chhattisgarh	–	–	–	–	0
6.	Goa	–	–	–	–	0
7.	Gujarat	4	4	1	2	11
8.	Haryana	–	2	–	–	2
9.	Himachal	6	3	1	1	11
10.	J&K	10	2	1	–	13
11.	Jharkhand	–	–	–	–	0
12.	Karnataka	3	–	–	-	3
13.	Kerala	–	–	–	–	0
14.	Madhya Pradesh	4	4	–	–	8
15.	Maharashtra	6	2	–	–	8
16.	Manipur	11	14	1	–	26
17.	Meghalaya	4	4	3	–	11
18.	Mizoram	5	–	–	–	5
19.	Nagaland	3	1	–	–	4
20.	Orissa	1	–	–	–	1
21.	Punjab	1	–	–	–	1
22.	Rajasthan	4	1	–	–	5
23.	Sikkim	1	1	–	–	2
24.	Tamil Nadu	-	1	–	–	1
25.	Tripura	3	1	–	–	4
26.	Uttaranchal	15	9	1	–	25
27.	Uttar Pradesh	2	2	–	–	4
28.	W Bengal	4	–	–	–	4

the same area. It measured 7.5 on the Richter scale and caused considerable damage in many parts of Assam. Thereafter, there was very little seismic activity for about 42 years till on 15 August, 1950, when there was a devastating earthquake measuring 8.5 on the Richter scale. This earthquake affected large parts of Assam and killed over 532 people. This earthquake was followed by intense seismic activity in the region for the next year or so. During this period there were over 70 earthquakes measuring 5 to 7 on the Richter scale. Then the seismic activity began to taper off. The last major one was in 1962 and it measured 6.5 on the Richter scale.

Assam also has a fairly active seismic history with 26 earthquakes with epicentre within the state. It has also been rocked by earthquakes originating in Arunachal Pradesh, Burma, Nagaland, Manipur, Mizoram and Meghalaya. Some of the epicentres have been in the foothills to the north. Some have been along the Bramhaputra valley. A number of them have been in the Cachar region. The earliest was the Cachar earthquake on 10 Jan., 1869 which measured 7.5 on the Richter scale. There were two strong earthquakes on 21 Jan., 1941 near Mangaldoi and 27 Jan., 1941 near Tezpur with both measuring 6.5 on the Richter scale. There was a major earthquake with epicentre near Dibrugarh on 26 August, 1950 which measured 7 on the Richter scale. There have been many earthquakes measuring over 6 on the Richter scale since then.

Bihar has had some seismic activity though it has been affected by major earthquakes in Nepal. The earthquake of 1934 has been the most devastating. Three minor earthquakes had their epicentre in the state. The last one was in the eighties.

Chattisgarh and Goa do not have any history of earthquakes.

Gujarat's history of earthquakes are mostly in the Rann of Kutch with 11 of its 13 earthquakes having their epicentre in this region. The other 2 were at Palanpur and Surat. The major earthquakes of Kutch have been as under :

- 16 June, 1819 measuring 8 on the Richter scale with epicentre near Bhuj. Over 2,000 people were killed.
- 19 to 25 April, 1845 measuring 6 on the Richter scale with epicentre near Lakhpat. 66 shocks were felt over the three days and some damage was caused.
- Three earthquakes were recorded on 29 Apr., 1864, 14 Jan., 1903 and 31 Oct., 1940.
- 21 July, 1956 measuring 7 on Richter scale at Anjar. It devastated the town and adjoining areas.
- January 2001 measuring 7.9 on Richter scale which has not only devastated Kutch but also caused severe damage as far away as Ahmedabad and Surat.

Haryana has little seismic activity. The state has recorded only two earthquake on 10 Oct., 1956 and 27 August, 1960. Both had their epicentre near Faridabad near the Moradabad fault. Both were over 6 on the Richter scale.

Himachal Pradesh has a long history of seismic activity. There have been 13 earthquakes with its epicentre in the state since 1856. It has also been rocked by earthquakes originating in Kashmir, Tibet, Uttaranchal and Western Nepal. The most devastating ones have been:

- 4 April, 1905 measuring 8 on the Richter scale with epicentre near Chamba. Its effect was felt over an area of 16,25,000 sq miles and over 20,000 people were killed.
- 22 July, 1945 measuring 6 on the Richter scale with epicentre near Chamba.
- 10 July, 1947 measuring 6 on the Richter scale with epicentre near Chamba. Caused serious damage at Bhadarwal where 73 aftershocks were felt.

J&K is also an earthquake sensitive state with about 13 earthquakes recorded since 1828. Of the 13, 3 were in Ladakh region in the Indus valley and 6 in the valley of the Jhelum. The most devastating ones have been :

- 6 June, 1828 measuring 6 on the Richter scale with epicentre near Srinagar. Over 100 people were killed and 1,200 houses destroyed.
- 30 May, 1885 measuring 7 on the Richter scale with epicentre near Baramula. The tremors were felt over an area of 11,000 sq miles and left over 6,000 people dead.
- 2 Sep., 1963 measuring 6 on the Richter scale with epicenter near Badgam south of Srinagar. 79 people were killed and over 400 were injured.
- 20 Feb., 1967 measuring 5.7 on the Richter scale with epicentre near Anantnag.

Jharkhand and Kerala have no significant history of earthquakes. Karnataka has recorded 3 minor earthquakes, the last one measured 5 on Richter scale near Bangalore on 7 Jan., 1916.

Madhya Pradesh has two major faults which are seismically active namely, the Tapi fault and the Narmada Rift. Most earthquakes are related to these. Some earthquakes have also been registered in the areas of Shivpuri and Guna. The major earthquakes in the area are :

- 2 June, 1927 measuring 6.5 on the Richter scale with epicentre near Rewa. The shocks were felt as far away as Dehrion Sone, Ranchi and some damage was caused at Jabalpur.
- 14 March 1938, 22 July 1945 measuring 6 on the Richter scale with epicentre near Indore. Damage was caused at Khandwa, Bhushawal, as also at Godra and Bharuch in Gujarat and Nashik in Maharashtra.

Maharashtra has seen some seismic activity along the Western Ghats at Koyna and Mahabaleshwar. Latur in Marathwada has also been seismically active. There are a number of faults around Mumbai but these are not active. The major earthquakes in Maharashtra have been:

- 10 Dec., 1967 measuring 6.5 on the Richter scale with epicentre near Koyna. The earthquake was preceded by tremors since 4 June, 1965 and aftershocks continued till 1970. Considerable damage was caused to villages in the subsequent areas.
- 30 Sep., 1993 measuring 6.3 on the Richter scale with epicentre near Killari. The earthquake resulted in over 8,000 people dead and 15,000 injured and totally destroyed about 80 villages. It rendered about 2,00,000 homeless.

Manipur has been a seismically active state. About 26 earthquakes have originated in Manipur since 1822. It has also been affected by earthquakes originating in nearby Myanmar. However, the damage to life and property has never been extensive. This is perhaps due to the type of buildings constructed with timber and bamboos which is prevalent in most of this area. The most serious earthquakes in the regions were:

- 3 Apr., 1822 measuring 6.5 on the Richter scale.
- 2 June, 1934 measuring 6.5 on the Richter scale.
- 11 May, 1940 measuring 6 on the Richter scale.
- 30 Apr., 1952 measuring 6 on the Richter scale.
- 1 July, 1957 measuring 7.5 on the Richter scale.

Meghalaya is also a seismically active state. The small state has experienced 11 major earthquakes since 1897. One of the worst earthquakes of all times originated here on 12 June, 1897. The quake measured 8.7 on the Richter scale and its tremors were felt over an area of 2,50,000 sq miles. Fissures and craters were abundant. However, the loss of life and property was not as devastating as could be expected in an earthquake of this magnitude. This was possibly due to type of houses constructed in the region. The other major earthquakes in the region were :

- 8 July, 1918 measuring 7.6 on the Richter scale with epicentre near Meghalaya-Bangladesh border.
- 9 Sep., 1923 measuring 7.1 on the Richter scale with epicentre on Meghalaya-Bangladesh border. The earthquake caused damage to property at Guwahati and Maymansing in Bangladesh.
- 2 July., 1930 measuring 7.1 on the Richter scale with epicenter near Tura in Garo Hills. It caused considerable damage to property at Guwahati, Jalpaiguri, Dhubri, Ranchi and Maymansing, Rangpur and Faridpur in Bangladesh.
- 19 and 21 June, 1963 measuring 6.2 on the Richter scale with epicentre in the Garo Hills.

Mizoram is not seismically very active. There have been 5 minor earthquakes in the state since 1818. Nagaland is also not earthquake prone. The state has seen 3 minor and one major earthquake measuring 6.5 on the Richter scale on 29 July, 1970.

Orissa is not earthquake prone and has seen only one earthquake in the last 200 years.

Punjab is also not earthquake prone. It has seen only one earthquake near Ferozepur.

Rajasthan is also not earthquake prone. Only 4 earthquakes have originated here. Three were in south Rajasthan near Mt. Abu and one was at Rajgarh in north Rajasthan, north of Churu.

Sikkim has experienced 2 earthquakes a major one on the western slopes of Kanchanjanga and other in West Sikkim. Tamil Nadu has experienced only one major earthquake measuring 6 on the Richter scale with epicentre near Coimbatore on 7 Jan., 1916.

Tripura has experienced two moderate earthquakes.

Uttaranchal is also a very seismically active state. The state has been the epicentre of over 22 earthquakes since year 1800. Uttarkashi, Almora and Dharchula areas have been most sensitive. It has also been affected by all major earthquakes of western Nepal. Though the earthquakes have been frequent, only one of more than 7 on the Richter scale has originated here. However, as most houses are made of stone in mud mortar, the damage is generally heavy. The major ones are listed below:

- 1803 measuring 6.5 on the Richter scale with epicentre in the Kumaon Hills. About 300 people were killed and many houses were destroyed.
- 28 August, 1916 measuring 7.5 on the Richter scale with epicentre near Dharchula. All houses at Dharchula collapsed and many died.
- 28 Dec., 1958 measuring 6.5 on the Richter scale with epicentre near Joshimath. It caused ground ruptures, gaping cracks and landslides in an area of about 100 sq miles and caused damage to several houses.

Uttar Pradesh has not been the origin of many earthquakes. Only about six having originated there in the last 200 years. But it has often been affected by earthquakes which had there origins in Nepal. Uttar Pradesh has three main faults, the Moradabad Fault, the Sohna Fault and the Lucknow Fault. The two significant earthquakes in the area were the Moradabad earthquake of 1956 and Bulandshahar earthquake of 1960. Both caused minor tremors in Delhi.

West Bengal is not seismically active. Only four minor ones originated in the state.

Seismic Zoning

"Seismicity" is a term variously used to describe the form and status of earthquake occurrence in a region. It takes into account the data on location of epicentre of the earthquakes, magnitude and maximum observed intensity, extent of the area in which the earthquake was felt, the time interval at which earthquakes of different magnitudes have occurred in the past and to express the potential earthquake risks accordingly.

The data on earthquake occurrence in the past indicates that the possibility of occurrence of an earthquake cannot be ruled out for any part of India. However, these occur with different intensity and frequency based on the tectonic units of the country. To prevent or reduce damage to buildings during earthquake, it is necessary to strengthen or design the buildings according to the expected intensity of the earthquakes, which may occur. In order to apply building codes properly in regions of different seismicity, it is necessary to prepare seismic zoning maps delineating areas of probable maximum intensity earthquakes.

Indian Standards Institute prepared the first Seismic Zoning Map of India in 1962. This was revised in 1966, 1970 and again in 1993. A rough map indicating the seismic zones is attached at the end of this chapter. The country has been divided into five seismic zones. The probable maximum magnitude of earthquakes that may occur in these zones are as under:

Seismic Zone	I	II	III	IV	V
Probable Maximum Magnitude	< 5	5 – 6	6 – 6.5	6.5 – 7	> 7

It needs to be appreciated that the Zone I does not mean that there cannot be any earthquake in the area. It only means that if an earthquake occurs in Zone I, the intensity of the earthquake is likely to be less than 5 on the Richter scale. The same logic applies to the other seismic zones. While designing buildings in any particular seismic zone, relevant building codes as applicable for that zone must be used. Earthquakes of intensity less than 6 on the Richter scale do not cause major damage.

Year Wise Occurrence

It is remarkable to note that the seismic activity has been very wide spread in some years. Let us see some examples.

1930 was the first devastating year. There was one measuring 5.5 on 11 May near Shimla. There was a major one in Meghalaya measuring 7.1 on 2 July and this was followed by 5 major aftershocks. Another two struck Manipur on 11 July and 22 Sep. The last one caused damage at Guwahati. One struck Andaman and Nicobar Islands on 7 July. On 25 June there was one at Shivpuri in Madhya Pradesh.

1937 was another devastating year. On 30 April there was an earthquake measuring 5.6 on the Richter scale in Tibet. On 20 Nov. and 11 December, there were two earthquakes measuring 5 and 6 at Uttarkashi. On 9 March there was one in Assam measuring about 5.5. This was followed by an earthquake of similar magnitude on 21 March at Manipur. Another measuring 6 struck Myanmar on 31 August and another followed on 9 Sep. on Manipur-Myanmar border. There was one on 12 March in the Andaman and Nicobar Islands.

1950 is perhaps the most remarkable in this respect. On 12 Aug. there was an earthquake in Himachal Pradesh measuring 5.5 on the Richter scale. On 12 Sep. there was an earthquake of similar magnitude in Sinkiang Province of China. On 15 Aug., 1950 there were two devastating earthquakes, one with its epicentre in Arunachal Pradesh measuring 8.5 on the Richter scale which killed 512 people and caused widespread destruction. The other measured 8 on the Richter scale and had its epicentre in Myanmar. On 26 August another quake measuring 7 on the Richter scale had its epicentre near Dibrugarh in Assam. There followed 90 earthquakes during 1950 covering the North Eastern states of Assam, Arunachal Pradesh and the neighbouring regions of Bhutan, Tibet and Myanmar. Three earthquakes of magnitude 6 on the Richter scale struck Andaman and Nicobar Islands on 15 Feb., 5 Apr. and 22 Dec., 1950. On 14 Jun., 1950 an earthquake measuring 5.6 on the Ritcher scale had its epicentre on the Pakistani side of the Rann of Kutch.

1956 is remembered for the devastating earthquake at Anjar near Bhuj in Gujarat on 21 July 1956. There was an earthquake of magnitude 6 near Tura in Meghalaya on 22 August, 1956. On 10th October an earthquake measuring 6.7 on the Richter scale struck Faridabad and caused minor damage to properties in Delhi.

1960s saw an earthquake of about 5.5 near Siliguri. One struck Itanagar in Arunachal on 26 May. Bhutan had an earthquake of magnitude 6.5 on 29 July. There were five earthquakes spread over from Jan. to Dec. in the Andaman and Nicobar Islands. An earthquake of intensity 6 struck Faridabad and caused some damage and injuries at Delhi.

On 11 January 1962, there was an earthquake at Raxaul on Nepal – India Border of magnitude between 5–6. There were two earthquakes of intensity of about 5.5 in Himachal Pradesh in 1962, the first was on 17 June near Chamba and the second on 15 September near Dalhousie. There was an earthquake in Uttaranchal on 14 July. Further east, there were two earthquakes in the North Eastern region. The first occurred on 22 September at Arunachal Pradesh. The second on 30th October at Nowgong, Assam. There were two in the Andaman and Nicobar Islands. The second which occurred on 16 Nov. was of 6.5 intensity and caused considerable damage to property at Port Blair, Mayabunder and Diglipur. On 1st September, an earthquake of intensity 5 rocked Palanpur area of Gujarat.

1965 was also an year of earthquakes in the country. On 29th January, there was one in North Kashmir. On 31st May there was one near Manali in Himachal Pradesh. There were two near Almora in Uttaranchal, one at Balrampur in Uttar Pradesh, one in eastern Nepal and four in Tibet. Further east there was one near Mongoldoi in Assam, two in Arunachal Pradesh and three in Myanmar. There were four in Andaman and Nicobar Islands and one in Kutch. The first foreshock at Koyna was also felt on 4th June 1965.

On Prediction of Earthquakes

Scientists are unanimous that it is not possible to predict the occurrence or magnitude of an earthquake. Astrologers make predictions. They take credit when their predictions come true and take a low profile when nothing happens. However, records of past earthquakes clearly indicate that major earthquakes are generally preceded by foreshocks and followed by aftershocks. The period of foreshocks and aftershocks can extend from two to three years. The periodicity of major earthquakes in earthquake prone areas appear to be once every 40 to 50 years.

Conclusion

The main conclusions we can draw from this chapter are as under:

- India has a long history of earthquakes. However, earthquakes, have occurred more often in some states and less often in others.

- Earthquakes can occur in any part of India. However, the intensity is likely to vary depending on the seismic zone in which its epicentre is located.
- Earthquakes of intensity of more than 6 on the Richter scale are liable to cause severe damage to life and property. Such earthquakes mostly occur in Seismic Zones III, IV or V.
- One earthquake of intensity more than 7 on the Richter scale seems to occur once every 30 to 50 years.
- Major earthquakes are preceded by foreshocks and followed by aftershocks. The duration of foreshocks and aftershocks can extend for two to three years. Some of these can be quite powerful.
- In some years, multiple earthquakes have taken place in many parts of the country.

Buildings must be designed as per the building codes applicable for the zone in which it is located. If the building is not so designed, try to get it suitably strengthened.

Record of Earthquakes—Kashmir and Himachal Pradesh

No.	Date	General Area	Magnitude	Remarks
1.	06.06.1828	Srinagar	6	100 killed, 1200 houses destroyed
2.	07.04.1856	Shimla	5	
3.	30.05.1885	Baramula	7	
4.	04.04.1905	Kangra	8	20,000 killed
5.	28.02.1906	Manali	7	Hundreds killed
6.	17.05.1917	Leh	6	
7.	11.11.1921	Leh	5.5	
8.	11.05.1930	Shimla	5.5	
9.	22.06.1945	Chamba	6	
10.	10.07.1947	West of Chamba	6	73 aftershocks
11.	12.08.1950	West of Chamba	5.5	
12.	12.09.1951	Kishtwar	5.5	
13.	17.06.1955	North Himachal	6	
14.	17.06.1962	Near Chamba	5.5	
15.	15.09.1962	Dalhausie	5.5	
16.	26.06.1963	Siachin	5.3	

17.	02.09.1963	Anantnag	5.3	
18.	07.07.1964	Gilgit	5.2	
19.	29.01.1965	North Kashmir	5.7	
20.	31.05.1965	East of Manali	5.3	
21.	16.03.1966	North Kashmir	5.2	
22.	20.10.1966	Ladakh	5.0	
23.	21.02.1967	Anantnag	5.1	
24.	11.02.1968	East of Leh	5.1	
25.	04.1986	Dharamshala	5.5	

Record of Earthquakes—Uttaranchal and Sikkim

No.	*Date*	*General Area*	*Magnitude*	*Remarks*
1.	1803	Kumaon	6.5	Many casualties
2.	05.03.1842	Mussoorie	5.5	
3.	11.04.1843	Almora	5	
4.	18.05.1852	North Sikkim	6.5	
5.	16.06.1902	Uttarkashi	6	
6.	13.06.1906	Uttarkashi	5.5	
7.	28.08.1916	Dharchula	7.5	All houses destroyed
8.	27.07.1926	Dharchula	6	
9.	08.10.1927	Dharchula	6	
10.	15.11.1937	Uttarkashi	6	
11.	04.06.1945	Almora	6	
12.	05.02.1949	Kedarnath	5.5	
13.	28.12.1958	Joshimath, UA	6.5	Major aftershock on 31.12
14.	21.08.1960	Siliguri	5.5	
15.	11.01.1962	Raxaul	5.5	
16.	14.07.1962	Gopeshwar, UA	5.5	
17.	30.01.1963	Almora, UA	5.5	
18.	27.11.1963	Gopeshwar, UA	5.1	
19.	30.08.1964	Sikkim	5.2	
20.	13.05.1965	Almora	5.1	
21.	18.11.1965	Almora, UA	5.2	
22.	27.06.1966	Dharchula	6.1	Five major aftershocks in next two days and four more in December
23.	05.01.1968	Bageshwar, UA	5.3	
24.	27.05.1968	Dharchula, UA	5.1	
25.	31.05.1968	Gopeshwar, UA	5.7	
26.	03.03.1969	Gopeshwar, UA	5.3	
27.	22.06.1969	Gopeshwar, UA	5.4	
28.	20.10.1991	Uttarkashi	6.6	

Record of Earthquakes—North East India

No.	Date	General Area	Magnitude	Remarks
1.	03.04.1822	Manipur-Myanmar	6.5	On the border
2.	08.07.1828	Manipur-Myanmar	6	On the border
3.	11.11.1842	Manipur-Myanmar	6	On the border
4.	19.12.1865	Manipur-Myanmar	6	On the border
5.	30.06.1868	Manipur-Myanmar	5	On the border
6.	10.01.1869	Cachar, Assam	7.5	
7.	12.06.1897	Shillong Plateau	8.7	One of the greatest earthquakes of all times.
8.	31.08.1906	Arunachal-Myanmar	7	On the border
9.	12.12.1908	Arunachal-Myanmar	7.5	On the border
10.	14.11.1915	Nowgong, Assam	5	
11.	08.07.1918	Tura, Meghalaya	7.6	
12.	09.09.1923	Goalpara, Assam	7.1	
13.	30.01.1924	Tripura	6	
14.	18.08.1926	Manipur	5.5	
15.	23.10.1926	Silchar, Assam	5.5	
16.	20.05.1926	Manipur	5.5	
17.	15.07.1927	Arunachal Pradesh	5.5	
18.	09.07.1928	Arunachal Pradesh	5.5	
19.	30.08.1928	Arunachal Pradesh	5.5	
20.	19.10.1929	Arunachal Pradesh	5.5	
21.	02.07.1930	Tura, Meghalaya	7.1	Major earthquake. Six major aftershocks in July
22.	11.07.1930	Manipur	5.5	
23.	22.09.1930	Manipur	6	
24.	06.03.1932	Silchar, Assam	5	
25.	24.03.1932	Tura, Meghalaya	5.5	
26.	27.03.1932	Silchar, Assam	5.5	
27.	09.11.1932	Nalbari, Assam	5.5	
28.	06.03.1933	Tura, Meghalaya	5.8	
29.	02.06.1934	Manipur	6.5	
30.	23.04.1935	Manipur	6	
31.	18.06.1936	Dhubri, Assam	5.5	
32.	09.03.1937	Mongoldoi, Assam	5.5	
33.	21.03.1937	Manipur	5.5	

34.	06.05.1938	Manipur	5.5	
35.	21.11.1938	Arunachal	6	
36.	27.05.1939	Manipur	6.75	
37.	13.02.1940	Karimnaganj, Tripura	5.5	
38.	11.05.1940	Manipur	6	
39.	21.01.1941	Mangaldah, Assam	6.75	
40.	27.01.1941	Tezpur, Assam	6.5	
41.	23.02.1941	Arunachal	5.5	
42.	06.09.1941	Mongoldoi, Assam	5.5	
43.	08.02.1943	Mongoldoi, Assam	5.5	
44.	23.10.1943	N. Lakhimpur, Assam	7.2	
45.	24.12.1944	Meghalaya	6	
46.	19.05.1945	Tura, Meghalaya	6	
47.	16.03.1946	Manipur	6	
48.	08.03.1947	Manipur	6	
49.	01.03.1948	Nagaland	5.5	
50.	10.12.1949	Coochbehar, WB	6	
51.	15.08.1950	Arunachal	8.5	One of the most destructive earthquakes. There were about 60 major aftershocks in August and another 60 in next 12 months
52.	07.04.1951	Meghalaya	6.8	
53.	30.04.1952	Manipur	6	
54.	26.05.1952	Arunachal	6	
55.	25.08.1952	Arunachal	6	
56.	07.11.1952	Manipur	6	
57.	04.05.1955	Arunachal	6	
58.	22.08.1956	Arunachal	6	
59.	01.07.1957	Manipur	7.5	
60.	04.01.1958	Arunachal	5.5	
61.	09.02.1958	Meghalaya	5	
62.	13.02.1958	Arunachal	5	
63.	14.02.1959	Arunachal	6	
64.	07.06.1959	Manipur	5.5	
65.	02.11.1959	Arunachal	5	
66.	26.05.1960	Arunachal	5	
67.	14.06.1961	Manipur	5.5	
68.	22.09.1962	Arunachal	6.25	
69.	30.10.1962	Nowgong, Assam	5.5	
70.	19.06.1963	Tura, Meghalaya	6.2	

71.	01.09.1964	Arunachal	5.7	
72.	21.10.1964	Arunachal	5.9	
73.	18.02.1965	Manipur	5.4	
74.	11.04.1965	Mongoldoi, Assam	5.1	
75.	06.10.1965	Arunachal	5.4	
76.	09.12.1965	Arunachal	5.3	
77.	24.02.1966	Guwahati, Assam	5.1	
78.	23.04.1966	Dhubri, Assam	5.1	
79.	26.06.1966	Nowgong, Assam	5	
80.	11.09.1966	Dibrugarh, Assam	5	
81.	26.09.1966	Bhalukpong, Assam	5.6	
82.	04.01.1967	Mizoram	5.4	
83.	08.02.1967	Mizoram	5.1	
84.	11.03.1967	Arunachal	5.3	
85.	14.03.1967	Arunachal	5.9	
86.	06.09.1967	Mizoram	5.1	
87.	15.09.1967	Arunachal	5.8	
88.	14.11.1967	Mizoram	5.1	
89.	23.01.1968	Nagaland	5	
90.	12.06.1968	Goalpara, Assam	5.5	
91.	18.08.1968	Goalpara, Assam	5.2	
92.	27.12.1968	Mizoram	5.2	
93.	25.01.1969	Tripura	5.4	
94.	26.02.1969	Tripura	5	
95.	28.04.1969	Nagaland	5.2	
96.	01.06.1969	Meghalaya	5	
97.	30.06.1969	Mongoldoi, Assam	5.1	
98.	30.09.1969	Manipur	5.4	
99.	19.02.1970	Jorhat, Assam	5.5	
100.	09.05.1970	Manipur	5	
101.	25.07.1970	Kishenganj, Bihar	5.2	
102.	29.07.1970	Nagaland	6.5	
103.	26.06.1971	Manipur	5.2	
104.	17.07.1971	Golaghat, Assam	5.3	
105.	04.1986	Mehgalaya	5.2	
105.	05.1987	Arunachal	5.7	
106.	02.1988	Arunachal	5.8	
107.	08.1988	Arunachal	7.2	

Record of Earthquakes–Indo Gangetic Plain and Rajasthan

No.	Date	General Area	Magnitude	Remarks
1.	13.06.1730	New Delhi	6.5	Last major earthquake at Delhi. Many ramparts and houses destroyed. Aftershocks continued for 4-5 months.
2.	04.06.1764	Banks of Ganges	6	Widespread death and destruction.
3.	01.09.1803	Probably Mathura	6.5	
4.	26.04.1848	Mt. Abu, Rajasthan	6	Mt. Abu town and Dilwara temple damaged.
5.	23.05.1866	Monghyr, Bihar	5.5	
6.	15.08.1906	Mt. Abu	5	
7.	09.04.1909	Mt. Abu	5	
8.	07.12.1911	Bardhaman, WB	5	
9.	06.11.1925	Lucknow, UP	5.5	
10.	25.06.1930	Shivpuri, MP	5.5	
11.	14.04.1934	Rajgarh, N. Rajasthan	5	
12.	29.08.1953	Balrampur, UP	6	
13.	10.10.1956	Faridabad, UP	6.7	Minor property damaged and some injured at Delhi.
14.	27.08.1960	Faridabad, Haryana	6	
15.	01.06.1965	Balrampur, UP	5.2	
16.	15.08.1966	Moradabad, UP	5.8	
17.	03.05.1968	Purulia, Bihar	5.6	
18.	24.10.1969	Mt. Abu, Rajasthan	5.3	
19.	14.10.1970	Ferozepur, Punjab	5.2	
20.	02.01.1988	Bihar	6.5	Extensive damage to life and property.

Record of Earthquakes—Kutch and Cambay

No.	Date	General Area	Magnitude	Remarks
1.	16.06.1819	Rann of Kutch	8	Near Bhuj, most destructive. Over 2000 killed
2.	13.08.1821	Anjar	5	
3.	25.04.1845	Lukpat	6	66 shocks were felt from 19th to 25th. Some destructive.
4.	29.04.1864	Rann of Kutch	6	
5.	14.01.1903	Rann of Kutch	6	
6.	31.10.1940	Rann of Kutch	6	

7.	14.06.1950	Rann of Kutch, Pak. Side	5.6	
8.	21.07.1956	Anjar, Kutch	7	Caused major destruction.
9.	01.09.1962	Palanpur, Gujarat	5	
10.	26.03.1965	Rann of Kutch	5.3	
11.	26.01.2001	Bhuj	7.9	Most destructive in recent times

The number recorded appears incorrect. Some of the foreshocks and aftershocks were not recorded or data is incomplete.

Record of Earthquakes Peninsular India

No.	*Date*	*General Area*	*Magnitude*	*Remarks*
1.	17.08.1764	Mahabaleshwar, Maharashtra	5.5	Felt up to Nashik
2.	29.01.1822	Vellore, Tamil Nadu	5	
3.	22.08.1828	In sea of Karwar, Karnataka	5	Felt up to Vengurla
4.	01.04.1843	Near Bellary, Karnataka	6	
5.	27.05.1846	Near Damoh, MP	6.5	
6.	18.11.1863	Indore, MP	5	
7.	08.02.1900	Coimbatore, TN	5	
8.	07.01.1916	Bangaluru	5	
9.	31.12.1926	Shivpuri, MP	5.5	
10.	02.06.1927	Rewa, MP	6.5	
11.	11.04.1929	Shivpuri, MP	5.5	
12.	14.03.1938	Bhushawal, Maharashtra	6	
13.	28.04.1951	Sea of Alibag	6	
14.	12.10.1959	Guntur, Andhra	5.5	
15.	08.05.1963	Sundergarh, Orissa	5.2	
16.	04.06.1965	Koyna, Maharashtra	5.4	First tremor at Koyna
17.	10.12.1967	Koyna	6.5	Main earthquake at Koyna. Ten major aftershocks.
18.	13.04.1969	Warangal, Andhra	6	Major aftershock next day.
19.	03.11.1969	Koyna	5.7	
20.	23.03.1970	Surat, Gujarat	5.4	
21.	25.09.1970	Koyna	5	
22.	26.09.1970	Mahabaleshwar	5.5	

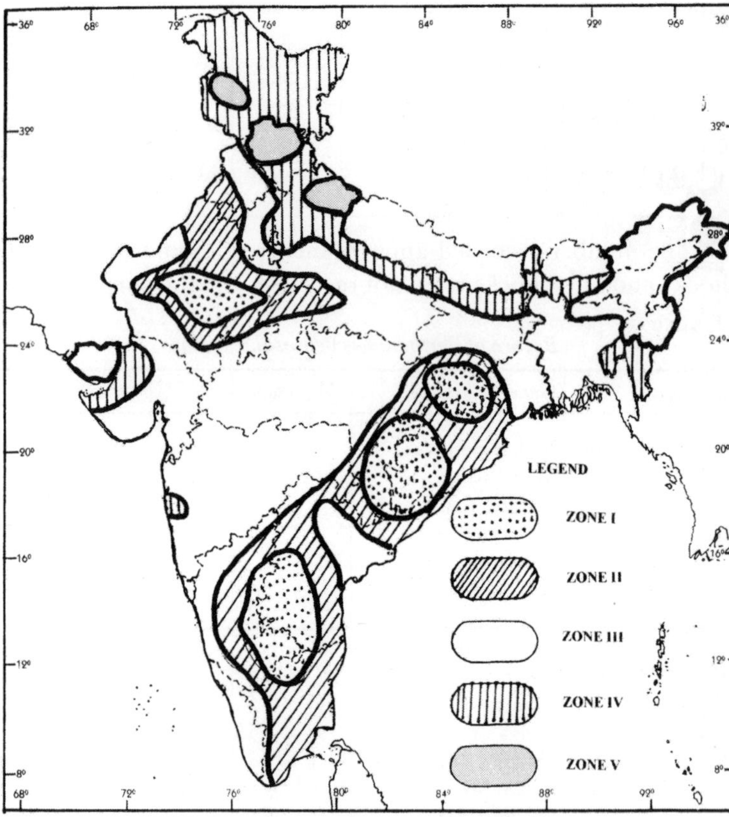

Seismic Zoning Map of India

2

CAUSES AND PREDICTION OF EARTHQUAKES

Cause of Earthquakes

The outer shell of the earth is called the lithosphere. The lithosphere is hard (something like the ice cap in the polar region) and cracked into a number segments called plates. These plates are given names like Eurasian Plate, Arabian plate, Indo-Australian Plate etc. The plates are in constant motion and in a particular direction. For example, the Indo-Australian Plate is moving northwards and pushing into the Eurasian Plate at a rate of a few centimeters per year. The crack between two plates is called a 'boundary fault'. The junction between the Indo-Australian and the Eurasian Plates runs through the Rann of Kutch to the Himalayas in North Kashmir-Tibet, along the general line of the Indo-Tibet and Nepal-Tibet borders up to Arunachal Pradesh and then turns south along the Indo-Myanmar border and through the Andaman and Nicobar Islands and then turns east and passes a hundred Kms south of Indonesia as shown in the map on the next page.

Pressure builds up when the plates push against each other, as the Indo–Australian plate does against the Eurasian plate. When the pressure becomes more than the edge of the plate can stand, a small part of the plate breaks or readjusts by moving up, down or sideways. This fracturing, crumbling or readjustment of the plate releases colossal amounts of energy and causes earthquakes.

The plates themselves have many cracks or faults. The main faults in the Indian Plate are shown in the Fault Map displayed at the end of this chapter. The description of the faults shown on the map is following:

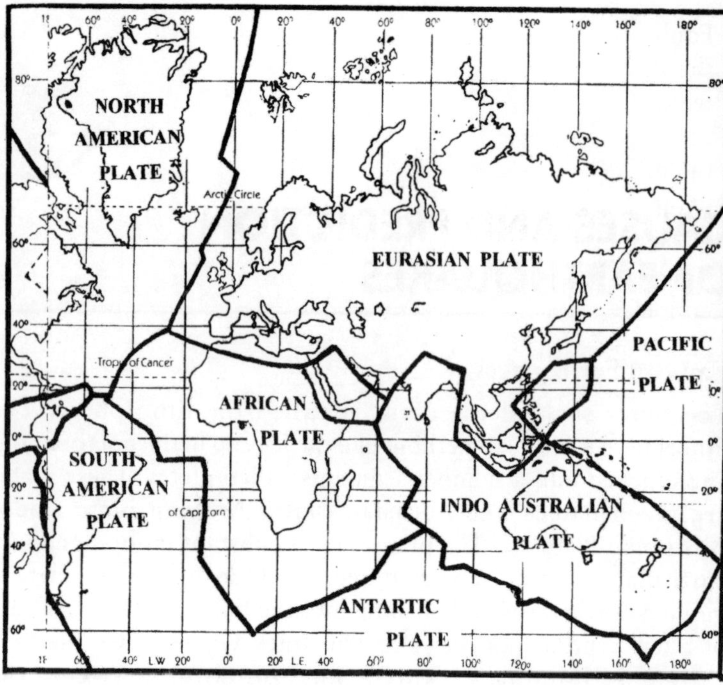

- **The Main Boundary Fault.** As explained earlier, this fault represents the junction between the Eurasian Plate and the Indo-Australian Plate. It crosses into Pakistan near Lakput in Kutch and re-enters India in Gilgit Region of Northern Pakistan Occupied Kashmir. Then it follows the Himalayan Ranges through Kashmir, Himachal Pradesh, Uttaranchal, Nepal, Sikkim, Bhutan, Arunachal Pradesh and then turns south through Nagaland, Manipur and Myanmar. It also passes through the Andaman and Nicobar Islands. The fault passes mostly through Seismic Zones V and IV. About 60% of earthquakes in India have occurred astride this fault.
- **The Counterthrust Fault.** This fault runs through Ladakh and Tibet on the northern side of the Himalayan Ranges. Many major earthquakes have taken place astride this fault. Many of these earthquakes have caused limited damage in Himachal Pradesh, Uttaranchal, Uttar Pradesh, Bihar, Assam and Arunachal Pradesh.

- **Fault No. 1.The Rann of Kutch Fault.** This fault runs along the Rann of Kutch and joins the Main Boundary Fault at its western end. This fault lies in Seismic Zone V and has been the cause of some of the major earthquakes of India including the Bhuj Earthquake of 2001.
- **Fault No. 2.** This fault extends from Palanpur in Gujarat to Sirohi in Rajasthan and generally marks the end of the Aravali Ranges. The fault lies in Seismic Zone III and has produced minor earthquakes affecting Mt. Abu and Palanpur.
- **Fault No. 3. The Great Boundary Fault.** This fault lies between the Aravali and the Vindhya Ranges. It has produced a few minor earthquakes near Shivpuri.
- **Fault No. 4. The Moradabad Fault.** This fault starts south of Moradabad and joins the Main Boundary Fault. It passes through Seismic Zone IV and have caused earthquakes affecting Delhi, Faridabad and Moradabad.
- **Fault No. 5. The Lucknow Fault.** This fault is near Lucknow and joins the Main Boundary Fault. The Bulandshahar earthquake was due to this fault.
- **Fault No. 6. The Patna Fault.** This fault also joins the Main Boundary Fault. Earthquakes affecting Monghyr and Madhupur were due to this fault.
- **Fault No. 7. The Dauki Fault.** This runs along the Meghalaya–Bangladesh border, through Cachar and joins the Great Boundary Fault. This fault is in Seismic Zone V and has been the cause of a number of major earthquakes.
- **Fault No. 8. The Narmada Fault.** This is one of the longest faults running from Jharkhand to Gujarat. It has been the cause of a number of minor earthquakes near Indore, Khandwa and Jabalpur.
- **Fault No. 9. The Tapti Fault.** This is a minor fault and has not produced any major earthquake.
- **Fault No. 10. The Cambay Fault.** This fault lies in the sea off the western coast of India. There have been a few major earthquakes on this fault. However, not much damage has been caused on land.

- **Fault No. 11.** This fault runs along the Western Ghats, south of Mumbai. They caused minor earthquakes at Mahabaleshwar and Koyna.
- **Fault No. 12.**This fault generally runs along the Krishna basin. No major earthquakes have been reported in this region.
- **Fault No. 13.** This fault generally runs along the Godavari basin. The Warangal earthquakes could be due to this fault.
- **Fault No. 14.** This fault starts in the Lakshwadeep and ends east of Coimbatore. No major earthquakes have been reported on this fault.
- **Fault No. 15.** This fault starts in the Lakshwadeep and generally demarcates the eastern limits of the Deccan Plateu. No major earthquakes have been reported on this fault.
- **Fault No. 16.** This fault generally marks the western limits of the Deccan Plateu. No major earthquakes have been reported on this fault.

Frictional pressures are also developing at these faults. When the pressures at these faults become more than that which the rocks at the fault can stand, the sides of the faults move relative to each other. This movement can be upwards, downwards and sideways. This movement and readjustments also cause earthquakes.

The faults that are located at the edges of the Plates, i.e. the Main Boundary Fault are more earthquake prone. This is borne out by the fact that most the earthquakes occur in this region. It will be seen from the data given in the previous chapter that most of the earthquakes that occurred in India between year 1800 to 1970 occurred in Kutch, Kashmir, Himachal Pradesh, Uttaranchal, Sikkim, Arunachal Pradesh, Nagaland and Manipur. All these states lie adjacent to the Boundary Fault between the Indo-Australian and the Eurasian Plates. The earthquakes in this region also tend to be of greater intensity. Most of the 7 plus magnitude earthquakes that have taken place in India during the above mentioned period, have taken place in this region. All 8 plus earthquakes have taken place in this region.

The faults within a plate also produce earthquakes. However, the earthquakes here are less frequent. The earthquakes produced by these faults are of lesser intensity. Most of these measure less than 6.5 on the Richter scale. Some of the major faults on the Indian Plate are shown in the Fault Map of India at the end of this chapter. It may kindly be noted that there are many other minor faults in the Indian Plate. In fact most rivers, creeks and hill ranges or cliffs are close to these minor faults. The earthquakes at Latur, Koyna and Mahabaleshwar in Maharashtra occurred on such minor faults.

The fracturing, crumbling and readjustments which occur at the faults due to the pressures built-up may occur at one instant. In such cases there is only one earthquake. But in other cases the fracturing, crumbling and readjustment occurs over a period of time. The action takes place bit by bit over a period of time with small fractures and readjustments preceding the major one. This produces foreshocks or smaller earthquakes before the main one. The period can extend up to two to three years as was experienced at Latur and Koyna Earthquakes.

Similarly, after a major earthquake, it may take some time before the fault again reaches a state of equilibrium. During this period small fracturing, crumbling and readjustment of the plates take place. These fracturing, crumbling and readjustments also produce smaller earthquakes, which are known as aftershocks. Immediately after the earthquake the frequency of aftershocks are more. Some of the aftershocks can be quite severe and measure 5 to 6 on the Richter scale. The frequency and intensity of aftershocks begin to taper off after a few months. Aftershocks can last for a period of two to three years. This was the case in the Assam Earthquake of 15 August 1950 and Koyna and Latur Earthquakes.

Earthquakes are due to upward and downward or sideways movement of the earths crust at the faults. This movement often results in changes in ground water levels. This was noticed at Latur and has been noticed in Gujarat after the present earthquake. Some of these changes may have taken place before the major earthquake and gone unreported.

Prediction of Earthquakes

Scientists are unanimous that it is not possible to predict the occurrence or magnitude of an earthquake. Astrologers make predictions. They take credit when their predictions come true and take a low profile when nothing happens. However, records of past earthquakes clearly indicate that major earthquakes are often preceded by foreshocks. Let us take some examples.

The major earthquake at Koyna measuring 6.5 on the Richter scale took place on 10 December 1967. However, the first earthquake in the same general area took place on 04 June 1965 and measured 5.4 on the Richter scale. This was followed by earthquakes of similar magnitude on 25 April 1967, 13 September 1967 and many smaller ones between June 1965 and December 1967.

In the Latur Earthquake of 30 September 1993, the case was similar. The area had been rocked by a number of small tremors for a period of two years before the disaster struck. Some changes in ground water levels were also reported.

The case of Pandhana and 24 other villages in Khandwa District of Madhya Pradesh have experienced 1,285 minor tremors since September 1998. The region is on the Narmada Rift, a well established fault which has seen a few earthquakes in the past. It is to be seen whether a major one will occur in the area.

The period of foreshocks and aftershocks can extend from two to three years. Hence, increase of seismic activity in any area, even if the tremors are of minor nature, does indicate the possibility of a major earthquake. However, the exact date of occurrence cannot be predicted.

Some believe that unexplained and sudden changes in level of ground water in wells and tube wells or flow in mountain stream is an indication of impending earthquakes.

The Chinese claim that they can predict earthquakes by studying the behaviours of birds and animals. Snakes come out of their holes in the ground before an earthquake. Cattle and horses refuse to enter their shelters without any apparent reason. The birds are also very restless.

Conclusion

The common man need not understand the reasons for occurrence of earthquakes. Neither does he need to be able to predict earthquakes. But living in fear can destroy the joy of living. What we need to do is to understand the risks that we face and take necessary precautions. We will deal with these issues in Part 2 and 3 of this book.

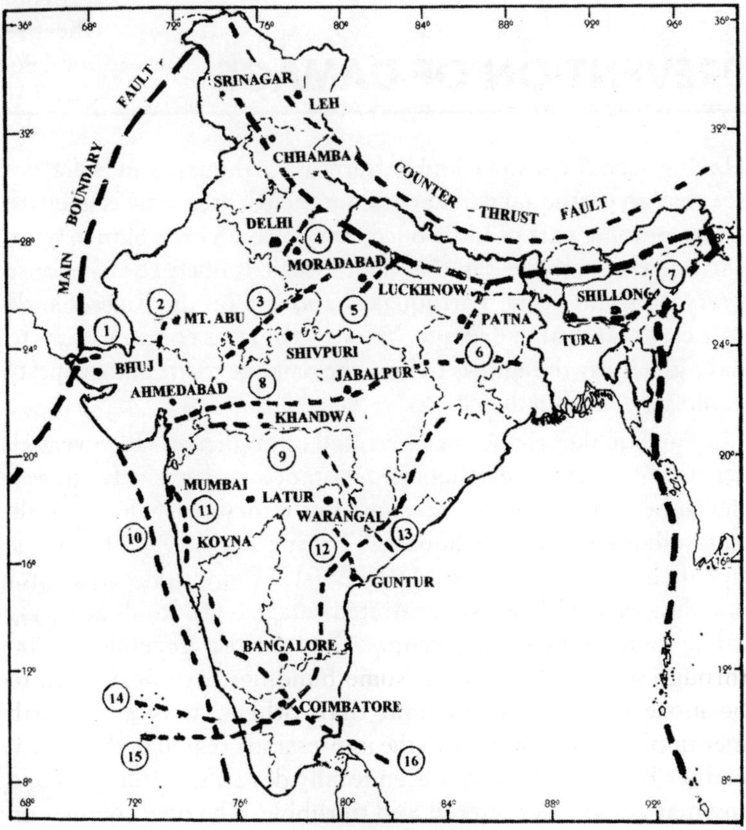

Fault Map of India

3

PREVENTION OF DAMAGE

Having seen the history and pattern of earthquakes in India, we see that an earthquake of about 6 on the Richter scale is likely to occur in some part of India once every 5 to 6 years. Similarly, an earthquake of above 7 in the Richter scale is likely to strike once every 40 to 50 years. Earthquakes above 6 on the Richter scale cause a great deal of damage. Yet no one in this country seems to have given any thought as to how the damage to life and property could have been reduced.

Earthquakes cannot be prevented or predicted. However, the science of design of earthquake resistant structures is fairly well developed. It is possible to construct structures, which would not collapse during earthquakes. During my survey of damage in Kutch from 5th to 11th March 2001, I found some structures standing even in the most devastated villages. The locals generally attribute it to divine intervention. But the fact generally is that through a matter of chance, some buildings are constructed to be more earthquake resistant than others. It is also worth mentioning that, contrary to the impression created in the media, neither Bhuj nor Anjar has been totally devastated. Only the old town at Anjar has been reduced to rubble. The new colonies on the periphery of Anjar have suffered negligible damage. At Bhuj also, most of the old houses and new multi-storied buildings have collapsed. Other newly constructed single and double storied buildings are largely intact. Only about 25 to 30% houses have collapsed at Bhuj.

Gujarat has a history of earthquakes. The epicentre of these earthquakes, are mostly in the Rann of Kutch. Yet, throughout my six days tour of Kutch, I could hardly identify five fully earthquake resistant buildings. After the visit, I am convinced

Standing Intact Among Ruins

that the loss of life and property would not have been even one hundredth of what has happened if the buildings in the region had been properly constructed. It is interesting to note that an earthquake of 6.8 on the Richter scale struck Seattle in the United States in February 2001. Only one person died, 250 were injured and not a single building totally collapsed.

Damage due to earthquakes can be largely prevented or at least reduced. What is necessary is to ensure that all new construction throughout the country is designed as per IS Codes on earthquake resistant construction. (Refer Bibliography). Existing buildings, particularly with stonewalls in mud mortar, need to be strengthened to make them earthquake resistant. This can be done by the following measures:

- **Improve Stability of Roof.** Remove stone, tile, or mud roofs and replace them with CGI sheet or asbestos sheet roofs or properly constructed RCC roofs. To reduce heat or cold, one can put tiles or thatch on top of the CGI sheets. One can also put a plastic sheet on the CGI sheet and lay a thin layer (50 mm) of mud mixed with cowdung and grain chaff on the plastic sheet.

- ***Strengthen Stone Walls Constructed in Mud Mortar.*** Insert 'headers' or 'through stones' in the stone walls. This should be done by carefully making one hole about 20 cm by 35 cm in the wall and inserting a through stone, approximately 15 cm × 30 cm × width of wall, in cement mortar. Such headers should be placed at 1.2 m spacing in the horizontal direction and 0.9 m spacing in the vertical direction. Those who can afford it should scour the joints in the mud mortar walls up to a depth of 6 mm and fix chicken wire mesh or 50 mm square mesh on the walls and plaster the walls in 1:6 cement mortar.
- **Create a Timber Framed Structure.** Support timber beams on timber posts so that the load of the roof does not come on to the stone walls in mud mortar. The timber posts should be connected to each other by timber 'ballies' or bamboos and properly braced to make a frame. The timber posts should be firmly fixed to the ground or floor.
- **Strengthen the Corners.** Carry out corner strengthening as explained in Chapter 14.
- **Strengthening of Load Bearing Buildings with Concrete Roofs.** Provide lintel bandage in lieu of band as explained in Chapter 14. The walls if constructed with stone in mud mortar should be strengthened as explained above. Also strengthen the corners as explained in Chapter 14.

The techniques of carrying out new construction and the points to remember regarding design and quality control are given in Chapter 15.

It is also time that we should stop depending on the Governments to solve all our problems. It is not possible for the Governments to educate the masses and guide them in making their houses more earthquake resistant. It is reported that 448 NGOs have come forward to provide rescue and relief to the earthquake victims in Gujarat. There are many other NGOs who are working in rural areas all over the country. I wish half of them could devote some time in guiding our rural masses in making their houses more earthquake resistant. That is perhaps the only way that large scale loss of lives and properties can be prevented during earthquakes.

Conclusion

Earthquakes can neither be predicted nor prevented. However, damage caused by earthquakes can be considerably reduced if dwellings are made earthquake resistant. It is possible to make existing houses more earthquake resistant. The techniques are simple and not very expensive. The NGOs could play a major role in this effort.

PART II

The Earthquake in Gujarat

4

THE EARTHQUAKE AND DAMAGE CAUSED

A massive earthquake struck Kutch region of Gujarat at about 0845 hours on 26 January 2001. It was the first Republic Day of the New Millennium and a holiday. School children were mostly at schools taking part in the flag hoisting celebrations. Most of the Ministers, Government Officials and policemen were assembled for the parades and flag hoisting ceremonies. In the towns and cities, families were in a festive mood. Late risers were enjoying a cup of tea in bed. Housewives were up and about doing their household chores. Vegetable and other vendors had set up their stalls in the markets and on footpaths. In the villages most of villagers were doing their daily chores. The farmers had left for the fields. Only the old, the women and the children were in the homes. Everything seemed normal and peaceful. Then suddenly all hell broke loose. For a minute or two the earth shook violently. Houses came tumbling down, trapping and killing thousands of people. Hundreds of thousand dreams were shattered. Thousands of families were bereaved. Thousands of crores worth of property and productivity were destroyed. The epicentre of the earthquake lay in the Rann a few kilometers north of village Lodai, about 20 km North East of Bhuj. A deep crater filled with water with bubbles coming out marked the spot. The tremors of the earthquake were felt in far away Delhi and Chennai.

The earthquake affected close to 40 million people in 21 districts of Gujarat and Rajasthan. Almost 9,000 villages in Gujarat and Rajasthan were affected. About 25,0,000 houses collapsed and many more suffered extensive damage. Official figures indicate that about 16,500 people died in the earthquake.

The Crater at Epicenter Near Lodai

The locals believe the number to be at least twice as many. Almost 150,000 people were injured. Kutch district was the worst affected. Some of its towns like Bhachau, Rapar and Vandh and around 100 villages were completely destroyed. Bhuj and Anjar towns suffered substantial damage. The Kutch district accounted for about 15,000 of the dead and 115,000 of the injured. Next was Rajkot District, which accounted for about 400 of the dead and 10,500 of the injured. The maximum devastation in the district was in and around the town of Morvi. Jamnagar district accounted for about a 120 dead and about 4,600 injured. Ahmedabad surprisingly accounted for about 750 deaths and about 4,000 injured. Damage and loss of life was also reported from the areas bordering the Rann of Kutch in the districts of Banaskantha, Mehsana and Surendra Nagar.

It is not my intention to render the heartbreaking tales of horrors and tragedies brought about by the disaster. Nor am I going to eulogize the instances of courage, sacrifice and human kinship that surfaced in the aftermath of the devastation. Most of you would have seen it all on the screens of your television sets or read about them in the magazines. The aim of this chapter is to record what happened in a manner so that we can find some method in the mayhem that occurred.

It is relevant to talk a little about the terrain in Kutch district. The central areas consist of hilly terrain. The hills are mostly of

Collapsed Multi-storied Building at Bachau

sand stone. All the towns except those on the cost like Bhuj, Bhachau and Rapar are located on hilly terrain. Sand stone is freely available and has been used extensively for construction in towns as well as the villages. The areas bordering the Rann are flat sandy plains.

Damage in Towns

I have visited most of the Kutch district and seen the damage in towns and villages of the district. Some photographs of the damage are a part of this book. The location of these cities, towns and villages and the extent of damage are shown on the map at the end of this chapter.

Ahmedabad. The distance of this city from the epicentre is about 400 km. About 131 multistoried buildings collapsed in the city killing about 750 people and injuring another 4,000 people. Interestingly, out of about 2,00,000 buildings, only about a 1,000 suffered damage. Quality of design and construction, rather than the intensity of the earthquake, seemed to be the cause of damage.

Bhuj. The town is located about 20 km from the epicentre. About 40 multistoried buildings and hundreds of old buildings with stone walls in mud mortar collapsed, killing and injuring thousands of people. However, hardly any of the single and double storied newly constructed buildings collapsed. Many of them showed cracks in walls. Less than 30% of the houses in the town have collapsed. Quality and type of construction rather than the intensity of the earthquake, seemed to be the cause of damage.

Anjar. Distance of the town from the epicentre is about 70 km. The congested old town consisting of buildings with stone walls in mud mortar have totally collapsed killing thousands. However, new localities around the old town have suffered little or no damage. The type of construction seems to have been the cause of the devastation. It is interesting to note that the same areas had been totally destroyed in the earthquake of 1956. However, the traditional building practices were followed during the reconstruction with disastrous results.

Adipur: The distance of the town from the epicentre is about 75 kms. About 20% houses, mostly built of stone walls in mud mortar, have collapsed.

Gandhidham: Distance of the town from the epicentre is about 80 km. Hardly 10% of the houses collapsed. Poor design, quality and type of construction rather than the intensity of the earthquake, seemed to be the cause of damage.

Kandla: Distance from the epicentre is about 100 km. The damage is similar to that at Gandhidham.

Bhachau: Distance of the town from the epicentre is about 50 km. 90% of the town has been reduced to rubble. Most of the buildings are concrete structures. In view of the distance from the epicentre, the extent of damage is puzzling. Poor design and quality of construction rather than the intensity of the earthquake, seemed to be the cause of damage.

Vandh: This town is about 5 km away from Bhachau. It has also suffered about 90% damage. This town had a large number of traditional houses of stone walls in mud mortar and tile roofs. Many newly constructed concrete houses have also collapsed.

Samkhiyali. This new town on the highway to Khandla is about 75 km from the epicentre. The town has suffered about 60% damage. The damage has been mostly to the traditional houses. One multistoried building has collapsed. Others have suffered damage.

Rapar. Distance of this town from the epicentre is about 80 km. The town has suffered about 60% damage. Almost all the traditional houses of stone walls in mud mortar and tile roofs have collapsed or suffered severe damage. Almost all recently constructed buildings in cement mortar are intact.

Mandovi: This town on the west coast is located at about 80 km from the epicentre. The damage is less than 10%. Only some old, traditional type houses have collapsed.

Morvi: Distance from the epicentre is about 100 km. The damage has been about 10%. Some new housing colonies have collapsed due to poor quality of construction. A number of old traditional type houses have either collapsed or suffered damage.

Wankaner: This town is located about 125 km from the epicentre. The damage has been about 10%. The old, traditional type houses in the old city and market area have been most affected.

Surendranagar: Distance of this town from the epicentre is about 175 km. The damage is less than 10%. The old, traditional type houses have been most affected.

Radhanpur. This town is about 170 km from the epicentre. Only minor cracks in some buildings were seen.

Surat: Distance of this town from the epicentre is about 600 km. A few multistoried buildings collapsed killing about 150 people. Poor design and quality of construction was the main cause of damage.

Damage in Villages

Most houses in the earthquake affected villages are constructed of stone walls in mud mortar. The houses have sloping tile roofs. Some of the prosperous villagers have built single and double storied houses with cement concrete roofs. Walls of cement concrete blocks, dressed stone and bricks are also seen in many villages. Most walls are plastered in cement mortar. Some of the houses in cement mortar have survived with damage to the walls. Rescue and relief took about 4 to 5 days to arrive.

Lodai: The village was close to the epicentre of the earthquake. Almost all buildings in the village collapsed.

Jawhar Nagar: This village is located about 10 km from the epicentre. It had been completely destroyed in the 1956

Earthquake and rebuilt by the Government. All the 215 houses constructed in the village in 1957 have been totally destroyed.

Dhamdhuka: This village, located about 30 km from the epicentre was also completely destroyed in the 1956 Earthquake and rebuilt by the Government. It has suffered over 95% damage.

Adoi: This village is located about 60 km from the epicentre and has suffered about 90% damage.

Ramvav: This village is located north of Adoi at about the same distance from the epicentre. It has suffered about 50% damage.

Ratnal: This village is located about 30 km from the epicentre on Bhuj Anjar Road and has suffered about 80% damage.

Sepa: This village close to Anjar is located about 40 km from the epicentre and has suffered about 50% damage.

Zinzuvada: This village is located on the edge of the Little Rann in Surendra Nagar district about 170 km from the epicentre and has suffered about 60% damage.

Analysis of Damage

It is interesting to note that the damage to buildings was not always proportionate to the distance of the town or village from the epicentre. For example, Bhachau, which is 50 km to the east of the epicentre, suffered almost total damage whereas Bhuj, which is at 20 km of the epicentre suffered only about 30% damage. The villages on Road Lodai—Bhuj, which is just 25 km away from the epicentre suffered only about 30% damage whereas Adoi located at about 60 km suffered almost total damage. Gandhidham town, which is much closer to the epicentre than Ahmedabad, suffered much less damage. Gandhinagar, which is just 25 km from Ahmedabad, suffered no damage at all. I have tried to analyze the reason behind this anomaly and find that the factors which influence the extent of damage are as under:

- Distance from the epicentre
- Type of building
- Height of building
- The foundation strata
- The quality of design and construction

It must be appreciated that the extent of damage does not depend on any single factor, but on the combination of all the factors.

Distance from the Epicentre

The intensity of the earthquake keeps reducing as the distance from the epicentre increases. Hence further the building from the epicentre, less are the extent of damage to buildings of similar type.

Type of Building

Buildings can be classified based on different criteria. These are discussed below:

- **Basic Pattern:** Fundamentally, the type of structures can be broadly divided into two types, load bearing structures and framed structures. Load bearing structures are those structures where the walls bear the load of the roof or the upper floors. These structures may have beams or trusses but all of them do not rest on columns. Most of them rest directly on the walls. In framed structures, a frame is first constructed with columns and beams or trusses. The frame is then filled with the walls. The walls do not carry the weight of the roof or the upper floors. The frames can be of timber, steel or reinforced cement concrete (RCC). In theory, framed structures are more earthquake resistant than load bearing structures. However,

Intact Framed Structure at Bhachau

it has to be ensured that the frames are complete, properly designed and in case of timber frames, that the joints are properly braced. Load bearing structures need bands to form a frame and to make them earthquake resistant. It was noticed that the extent of damage to a building did not depend on whether the building was a load bearing structure or a framed structure. The extent of damage depended on whether the building was correctly designed and whether proper quality control was exercised during construction.

- **Type of Walls:** The walls of the building can be of bricks, stones or concrete blocks. Stone walls in mud mortar are most vulnerable. In the absence of use of 'headers', they tend to split down the middle and total collapse can be expected in case of earthquakes. 90% of buildings collapse in the Gujarat earthquake was due to stone walls in mud mortar. Walls in cement mortar can crack. But total collapse is unlikely. Brick walls are more stable than concrete or stone blocks. This is because bricks have frogs, which make the bond between courses stronger. The bricks in adjoining courses are laid at right angles to each other. This also seems to increase stability.
- **Type of Roof:** The roofs can be of CGI sheets or RCC slabs, Mangalore tile, stone slab or slate and mud. Light roofs like those of CGI sheets are the safest. Mangalore tiles are unstable and are disturbed easily during earthquakes. They injure but

Mangalore Tile Roof Damaged in the Earthquake

Collapsed RCC Columns and Roofs

do not kill. Thousands of houses with Mangalore tiles did not collapse but lost the roofs. Stone and slate slabs are heavy and dangerous. They kill if the collapse. In theory, RCC roofs are safe. However, they are often poorly designed and constructed and collapse. They are heavy and can kill.

Height of Building

The taller the building, more vulnerable it is during earthquakes. In the towns and when built in cement mortar, single and double

Collapsed Ground Floor, Multi-storied Building

Collapsed Roadside Hotel

storied buildings rarely collapsed. Multi storied buildings suffered the maximum damage. This was very much in evidence at a majority of the towns where most of the single or double storied houses survived the earthquake but almost all multi-storied buildings collapsed. However, when properly designed and constructed, they can survive earthquakes.

Foundation Strata

The foundation strata has a major bearing on the extent of damage. Buildings and colonies located on hard or rocky strata suffer less damage during earthquakes.

Quality of Design and Construction

- **Earthquake Resistant Design:** Very few buildings in Kutch have been built to earthquake resistant specification. This is in spite of Kutch being located in Seismic Zone V. Even buildings, reconstructed after the major earthquake of 1956, which destroyed Anjar, were not made earthquake resistant. Even Government buildings of the district administration, state electricity boards, police housing, defence establishments, telecommunication department are not designed to be earthquake resistant.
- **Parking on Stilts:** This appear to be unsafe in earthquake prone areas. Many buildings were seen where the stilts had collapsed

Typical Stone Wall in Mud Mortar

but the rest of the building appeared reasonably intact. For some reason, possibly due to incorrect structural design, the ground-floor of many multistoried building have collapsed.

- **Shape of the Structure:** Buildings of irregular shape or with large cantilever projections appear more vulnerable.
- **Quality of Designs and Construction:** This seems to be the main reason for the collapse of multistoried buildings at Ahmedabad, Surat, Bhuj, Bhachau and the other towns. The reinforcements at the junctions of columns and beams, particularly at ground floor levels appeared to be inadequate and incorrectly placed. Many frames were incomplete with plinth beams and some of the other beams missing. In many cases, the quality of mortar and concrete was found to be poor.

Conclusion

The massive damage to life and property that resulted from the earthquake was primarily due to complete lack of earthquake resistant construction. Kutch is located in Seismic Zone V. It has been rocked by a large number of earthquakes. Yet earthquake resistant construction is totally unknown in the region.

In the villages, the type of construction was largely responsible for the damage. As seen in earlier earthquakes at Latur and Uttarkashi, stone walls in mud mortar without use of 'headers' are extremely vulnerable. These kind of walls are responsible for 90% of the damage. When supporting concrete roofs, the collapse of these walls brought down the roof and killed the occupants. Tiled roofs were responsible for the balance 10% of damage.

In cities, the damage was entirely due to poor quality of design and construction. The practice of building structures close to each other was also the cause of damage, particularly at Anjar. Collapsing buildings pulled adjacent ones down.

It is a matter of shame that Government buildings, which are supposed to be designed by competent structural engineers and constructed by competent contractors, suffered such massive damage. It points to gross apathy and corruption.

In Kutch, the history of disaster has been repeated again. If due care is not taken during the reconstruction and standing houses are not effectively repaired and made earthquake resistant, the history of disaster will be repeated again.

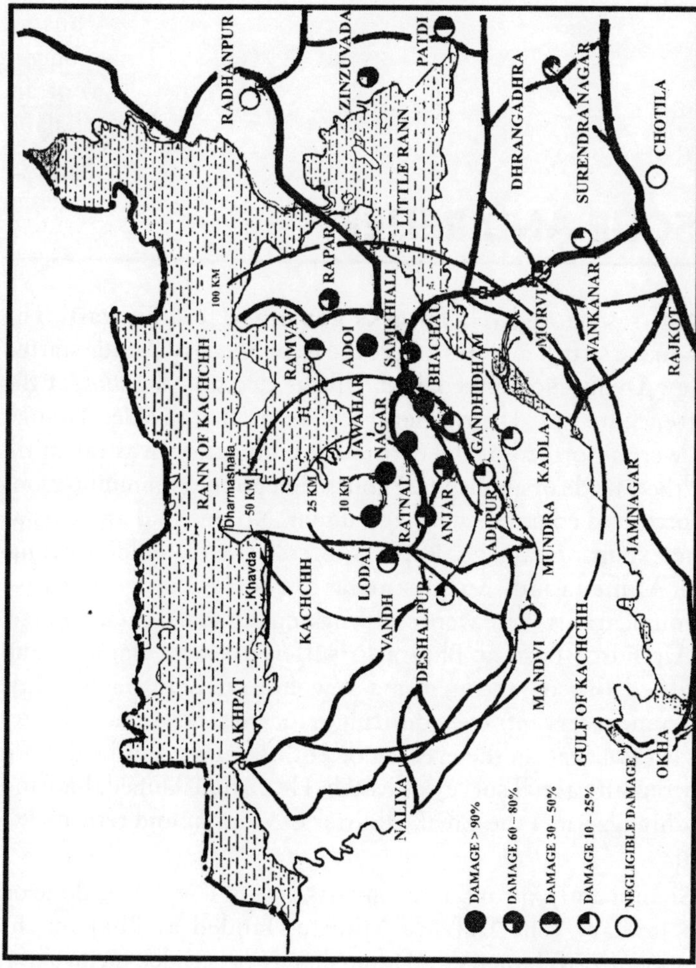

Damage Map of Kachchh and Adjoining Areas

5

RESCUE AND RELIEF

Official rescue and relief works got off to a slow start. The earthquake struck minutes before the Republic Day Parade started at New Delhi and Ahmedabad. All the decision makers, at the Centre, State and District levels, were at the parade. Though they were informed of the earthquake, no action was taken till after the parade or about 1100 hours. Telephone communications to Kutch had completely broken down. No one had any clue as to the extent of damage. Reports of collapse of buildings came from Ahmedabad, Surat and a few other cities because communications there were still working. To make things worse, the Air Force base at Bhuj also suffered severe damage and remained out of action for a few hours. Being a holiday, Government servants were difficult to locate. The Home Minister did visit Gujarat on the evening of 26th January. But he did not carry out any aerial survey of Kutch. He visited Ahmedabad and Gandhinagar, met the Chief Minister of Gujarat and returned to Delhi.

Significantly, no aerial reconnaissance of Kutch was done on 26th January. The Defence Minister landed at Bhuj on the morning of 27th January and immediately proceeded on an aerial reconnaissance of the earthquake affected areas of Kutch. The true picture of the devastation began to unravel on his return and as the TV cameras of the media reached the sites.

The National Highway 8A to Kandla port from Ahmedabad via Morvi had been blocked. The bridge across the Little Rann near Surajbari had been damaged by the earthquake. The broad gauge railway line to Bhuj along with about 40 railway stations were damaged. Kutch was thus only accessible from the rest of the country on National Highway 15 through Radhanpur. For some reason this road was not used by rescue columns from Ahmedabad. The main link to Kutch for first three days was by air.

Rescue Operations

Immediate rescue and relief was left to the initiative and resourcefulness of the local people. They had no expertise or wherewithal. They tried to break through the concrete debris with primitive tools like sledge hammers and crowbars and reach those trapped under the debris. Cranes and JCBs were scarce. The approaches to some of the sites were blocked by the debris. Army and Paramilitary Forces started reaching the collapsed buildings at Ahmedabad in the afternoon. With their arrival, the rescue efforts picked up. But without heavy equipment to remove debris quickly, the progress was agonizingly slow. In spite of the difficulties, many lives were saved. The first Army Engineer units arrived at Bhuj by air on 26th itself. Others arrived by road on 27th and joined the rescue efforts.

The Government of India took its time to accept offers of assistance from foreign countries. Thus the expert rescue teams began to arrive at Ahmedabad on the evening of 27th January. There were further gaffes. Customs clearances took time. There was no organized plan to coordinate their deployment and arrange their transport. The teams meant for Bhuj were dispatched by bus rather than air. Thus valuable time was lost. However, in spite of the various problems, the foreign rescue teams affected many miraculous rescues. They also demonstrated the capabilities of the state of art rescue equipment and utility of sniffer dogs in rescue operations.

The rescue efforts in Ahmedabad and other cities and towns outside Kutch got into full gear after about 48 hours with the arrival of heavy equipment in adequate numbers. But by then it was too late for many. Officially, the rescue effort was called off on the fifth day. Foreign rescue teams were surprised at the decision as their experience indicated that live survivors could still be expected. However, rescue efforts continued at some locations and some more lives were saved. After that it was time to clear the debris and dig out the dead.

The situation in some of the small towns in Kutch district was pathetic. Bhuj, Anjar, Bhachau, Rapar, Vandh were largely devastated. Organized rescue efforts reached these towns in the form of Army units on the morning of 27th. Specialized teams from foreign countries arrived only on 29th. Even then, it was too little for the magnitude of the task at hand. Paucity of heavy specialized equipment like concrete cutters, cranes and excavators severely affected the rescue efforts. The local people and local NGOs like RSS did whatever they could and saved many lives. Industrial houses and contractors provided whatever assistance they could. Some miraculous survivals were recorded.

The situation in the 900 odd villages of Kutch district was even more pathetic. No organized rescue effort reached most of the villages. The local people and some NGOs who had arrived did whatever they could. Fortunately, majority of the roofs were of tiles and many of the people were outside the houses. Hence casualties were less than what could be expected.

Relief

The relief effort was better. There were no major problems in Ahmedabad and the other cities. Problems were acute in Kutch. The civil hospital at Bhuj had collapsed during the earthquake killing many of the patients and staff. Initial medical aid was provided by the Air Force hospital at Bhuj. An Army Field Hospital was flown in on 28th January. Another Army Field Hospital, a Japanese field hospital and a hospital ship of the Indian Navy soon joined in to provide medical aid. Many medical teams were flown in. Emergency surgery was carried out round the clock in tents and even in the open. Serious patients were flown out to Mumbai, Pune, Delhi and some of the other cities.

The nation responded magnificently to the calamity. Food and other aid flew in fast. The NGOs and local people from all walks of life played a praiseworthy part in providing relief. Shelter and blankets however remained a problem for weeks. Drinking water and sanitation were also problems. In a couple of days it became a problem of plenty. Relief material piled up in heaps at Bhuj. No effective distribution system was available for the same. Nothing reached many villages for days.

Disposal of the dead posed problems. There was a paucity of firewood for cremations. Tool for digging graves were in short supply. Identification and certification of the dead also posed problems in the initial stages. The Army, BSF and NGOs, particularly RSS and some locals played a major role in this difficult task.

Coordination of rescue and relief effort remained a major problem for more than a month. There was no central authority to direct the rescue and relief work. Army, Paramilitary Forces and NGOs selected their areas of operation and did what they considered appropriate. This resulted in over concentration in some areas and paucity of effort in others. Allocation of specialized equipment also posed problems. As far as distribution of the relief supplies was concerned, the situation was pathetic. No one knew what was required and where. There was no organization to take charge of what arrived and to send it to where it was most needed.

By the time I visited the area a month and a half after the earthquake, relief appeared to be well organized. The responsibility of looking after relief at towns and villages had been distributed to NGOs, industrial houses, Army and other Central Government Organization. Tented camps had been established. Water supply and supply of rations had been organized. Make shift schools and medical centers had been set up. There were signs of revival of economic activity.

Some Points to Ponder

In case of major disasters like earthquakes and cyclones, the local people including the Government machinery at district level are badly affected. Most people are in a state of shock and dismayed by personal losses. Morale is low. It is therefore unfair to expect the district administration to be effective enough to mount large scale relief and rescue operations from within their own resources. The immediate rescue and relief has to be provided by Army, Paramilitary Forces and NGOs who move in from outside the disaster zone. The best that the State Government can do is to have contingency plans by which resources from outside the district move to the disaster zone and take over the administration.

Coordination was a major problem. There has to be some authority to control activities and to direct the rescue and relief work. This should be the responsibility of the State Government. However, as they are not effective in the initial stages, this task has to be done by some other organization. Army is the only organization, which has the capability to take on this task.

Receipt, storage and distribution of relief supplies require trained and responsible manpower, transport and communications. There must be staff at the villages and towns where the supplies are required, to intimate the requirements of the village or town to the control headquarters and to distribute it in an orderly manner when the supplies arrive. There must be staff at the control headquarters to receive the demands, divide the supplies received and arrange to deliver the stores where they are required. There are requirements of unloading and loading staff and guides. This is a massive task and only the Army has the wherewithal to handle the task. In this task, they would require the assistance of NGOs, local people and the State Government.

Rescue of people trapped in rubble is a specialized task. It requires specialized equipment like concrete cutters, cranes, sniffer dogs and trained experts. The capability of Indian organizations and NGOs was found to be inadequate. This aspect must be attended to.

In rescue and relief operations, time is of essence. Mounting a large scale rescue and relief operation requires detailed contingency planning and resources. Only the Ministry of Defence seemed to have any contingency plan in existence. This is borne out by the fact that they were able to air lift about 2,000 Army personnel into the disaster area within 24 hours and deploy over 20,000 within about 72 hours. But the tasks allotted to them were limited to rescue and relief and not to take control of rescue and relief operations. No effective disaster management plan at the Central Government and State Government level was in evidence.

Conclusion

The rescue and relief works carried out after the earthquake were largely ad hoc. There was little coordination amongst agencies. The whole effort was left to the initiative and resourcefulness of the local people and the organizations partaking in the effort. To be more effective, such effort needs to be better coordinated, organized and controlled. Central and State Governments must have a workable disaster management plan.

6

ROLE OF THE ARMED FORCES AND PARAMILITARY FORCES

The Armed Forces have always played a major role in rescue and relief after all major disasters. The Gujarat Earthquake was no exception. This was one of the rare cases where all the three services played a role. Over the years, the Paramilitary Forces have also been playing an increasing role in the national affairs. The BSF participated in strength in the rescue and relief operations.

The Air Force

The Air Force was the first to enter the fray. The Air Force Station at Bhuj was damaged by the earthquake. The control tower, some of the office buildings and married quarters were damaged. About a hundred including servicemen and families died in the earthquake. An equivalent number were injured. Fortunately, the runway was not damaged. A temporary control tower was established and the airport remained operational day and night. In the first five days after the disaster, the Air Force flew in 232 sorties of relief supplies. The Air Force also air lifted Army units to Bhuj from all over the country. Engineer regiments were flown in from Pune and Bhatinda.

The Air Force was also the first to provide medical aid in Kutch District. With the civil hospital at Bhuj completely destroyed, the Air Force Hospital was the only surviving hospital at Bhuj. Even this hospital had been damaged by the earthquake. The hospital was shifted into tented accommodation in the open and began treating the injured within hours.

The Navy

The Navy also got into the rescue and relief act. It ferried and provided drinking water. It brought in relief stores from Mumbai. It set up two hospital ships and provided treatment to thousands of the injured.

The Kandla Port suffered moderate damage during the earthquake. The Kandla city also suffered some damage. Most of the civilian staff of the port abandoned their post and disappeared. The Navy took over and operated the Kandla Port till the Port Authorities were in a position to take charge.

The Army

The Army played a major role in the rescue and relief operations. By the evening of 26th January, the Army had deployed about 5,000 troops at Ahmedabad and Bhuj. This number swelled to over 22,500 within the next five days. They were deployed at Anjar, Bhachau, Bhuj, Adoi and many other villages. They worked round the clock, rescuing people, digging out dead bodies and disposing them, demolishing dangerous buildings, distributing relief material and protecting the ruins from vandalism by anti-social elements. They also played a major part in clearing the rubble. The army was also the first to reach out to the villages and find out the extent of damage and casualties. This information was later used by civil administration. The Army also assisted government department in reestablishing themselves by constructing temporary accommodation for them. They assisted banks and the effected people in recovering cash, jewelry and possessions from the collapsed and damaged buildings.

The Army also played a major role in providing medical aid. It moved in two Field Ambulance Units and set-up two hospitals at Bhuj airfield. They also provided logistic support to the medical teams who came from all over the country and enabled them to provide medical treatment to thousands of the injured.

The Army continued to play a major role in the relief operations. They have adopted many villages and set-up tented camps for the homeless. They assisted in setting up water treatment and supply systems and improving sanitary facilities.

The task of demolishing dangerous damaged buildings was continuing even after 45 days. The buildings to be demolished are notified by the district administration. These are then carefully brought down. The Army had demolished over 50 such buildings. More are being notified.

The BSF

About 3,000 Paramilitary personnel, mostly of the BSF and some personnel of the Rapid Action Force also took part in the rescue and relief operations. The BSF took charge of Rapar Taluka and did a commendable job both during rescue and relief operations.

What More could have been Done

This is not meant to be a criticism of the Army. They played a major and commendable role in the rescue and relief operations. Perhaps they could have done more. Possibly no one thought about the possibilities. It is possible that these aspects were not in the disaster management contingency plan based on which the Army deployed.

Firstly, the Army has its own helicopters. With the help of these helicopters, the Army could have carried out aerial reconnaissance of the affected areas on 26th January itself and a clearer picture of the extent of the disaster and the damage to surface communications would have emerged. This would have reduced the response time required to mount the rescue and relief operation. An aerial reconnaissance of Ahmedabad on the morning of 26th January would have clearly indicated the extent of damage and the best routes to reach the sites of the collapsed buildings.

Secondly, the Army is capable of setting up telephone and radio communications in a matter of hours. For some reason, this capability was never used.

Thirdly, the Army should have been made responsible for finding out the extent of the devastation, particularly in the villages. They could have sent out reconnaissance parties to fan out to the villages and sent in much needed information for planning rescue and relief operation.

Conclusion

The Armed Forces are most suited for rescue and relief during disasters. Their units are self-contained in terms of administration and can move at short notice. They are used to operating in crisis situations. They should take charge of the rescue and relief operations till the State Government administrative machinery can be mobilized and become operational.

7

ROLE OF NGOs AND INDUSTRIAL HOUSES

The NGOs always play a very significant role in providing rescue and relief after disasters. They have the related experience and dedicated volunteers who can react quickly to these kinds of situations. The Gujarat Earthquake was no different. 448 NGOs, big and small, local and from all parts of India, deployed about 7,500 volunteers in the rescue and relief effort. There were Tibetans from Dalai Lama's followers at Dharmshala. There were NRI organizations. There were groups from Delhi and Punjab. There were spiritual NGOs, NGOs specializing in rural development and child care. As a group, they were only second to the Army in terms of the effort deployed.

NGOs

The local NGOs like the RSS got into the act within hours of the earthquake. Those from outside Kutch started arriving from the second day onwards. Unfortunately, there was no one to tell them where to go. So the NGOs selected their village or town and started the rescue and relief works.

The NGOs played a significant role in the rescue effort. Their volunteers dug out survivors from the debris and saved many lives. This apart, the NGOs carried out the following tasks:

- Provided medical assistance and arranged for evacuation of casualties.
- Procured and provided cranes, bulldozers and other heavy equipment for removal of debris.
- Distributed cooked food, water and other relief supplies.

- Assisted in recovery and disposal of dead bodies. In this the RSS played a very major role.
- Provided shelter to the homeless.
- Provided a psychological healing touch to the survivors.

Many of the NGOs are still active in the earthquake affected areas. They have adopted villages, set-up tented camps for the homeless. They have organized water supply and electrification. They provide ration and cooked food. They have set-up temporary schools, medical aid centres and recreation facilities for the children. Each NGO has worked to the best of their ability and resources. Without their participation, providing relief would have been almost impossible.

Some of the NGOs propose to take part in the rehabilitation process. They played a major role in rehabilitation of the victims of the Latur Earthquake. It is certain that they would do the same in Kutch.

Industrial Houses

The industrial houses with their units in Kutch and adjoining areas also played a significant role in the relief and rescue operations. They deployed their manpower and heavy equipment for rescue and removal of debris. Many industrial houses like Reliance, Tatas, Larsen and Tubros, Shell Gas etc. have adopted villages and towns and set-up tented camps for the homeless. They are also catering to the basic needs of the people. Some of the industrial houses also plan to participate in the rebuilding of Kutch.

Conclusion

The contribution of NGOs in disaster management has always been significant and will be so in all future disasters. They have the expertise and ability. What is lacking is detailed knowledge about their capabilities and strengths. Detailed contingency planning for their deployment and coordination of their effort will go a long way to get the best out of them.

8

ROLE OF THE GOVERNMENT

The role of the Gujarat Government in rescue and relief operations has been the subject of much criticism in the media. Part of the criticism is justified. But there are some mitigating factors, which need to be understood. The main allegations against the Government are:

- The Government was totally unprepared for a disaster.
- Key officials were either not available or not effective.
- Government failed to contribute in any way to the rescue operations.
- The Government was unable to coordinate distribution of relief supplies
- Government is not responsive to the plight of the villages.
- Government had allowed illegal and unsafe construction.

Was the Gujarat Government Totally Unprepared for a Disaster

It is perhaps true that the Gujarat Government was not prepared for the disaster. There is no evidence to suggest that the Government has a viable and implementable Disaster Mitigation and Disaster Management Plan. If it has one, it should publish a white paper on the subject. It is also common knowledge that the satellite based communication system installed after the last cyclone was non-operational. Instructions from the state capital to ensure the operational readiness of the system were mostly ignored.

While accepting the above, it must be realized that it is not cost effective to have dedicated resources for disaster management. Major earthquakes occur once in 40 to 50 years. Cyclones and floods are more frequent. But then these calamities can not be predicted. Hence it is easier to mobilize resources to tackle such disasters. Earthquakes strike without warning. Hence they are the most difficult to deal with. And when such a disaster strikes at a long weekend, it becomes even more difficult to deal with them.

The best that any Government can do is to have a comprehensive and implementable disaster management plan.

Were Key Officials Either not Available or not Effective

This was also a fact. The administrative authorities at Ahmedabad and Surat were available. But they were not clear as to what was expected of them. So they tended to remain mute spectators. In the absence of a clear disaster management plan, this is bound to happen.

The situation at Kutch was different. The district administration itself was affected by the disaster. Many lost their family members in the earthquake. Many had lost their homes. The infrastructure like electric supply and telecommunications was down. The staff was in a state of shock. Their inability to react to the disaster in an organized manner is understandable. What is perhaps inexplicable is the inability of the Government to induct necessary personnel from outside the district to take control. The absence of a clear contingency plan was responsible for the state of affairs.

Did the Government Fail to Contribute in any Way in the Rescue Operations

It appeared to be so. But it needs to be understood that officials of the State Governments, by themselves, can do nothing in case of disasters like earthquakes. More than 30, 000 members of the Armed Forces and volunteers from NGOs assisted by an equal number of local people worked round the clock to cope with the situation. There is no way that any State Government can mobilize such resources in a matter of hours. The fact is that there is little that a State Government can do by itself in case of a major disaster. It has to depend on the armed forces, NGOs and the local people to carryout the rescue and relief. The best that any State Government can do is to restore the infrastructure as early as possible and to act as a facilitator for the rescue and relief operations.

The main role of the Government is to coordinate the rescue and relief operations and to get outside assistance as early as possible. This coordination was conspicuous by its absence at least in the initial cases. This was because there was no plan or resources for carrying out such coordination.

Was the Government unable to Coordinate Distribution of Relief Supply

If the media reports are to be believed, this was true. In fact this had to be true. The first requirement of effective distribution of supplies is to know what is required where. The second requirement is to find the places and deliver the relief supply. The third requirement is to have an organization to coordinate and control receipt and distribution and monitor progress. None of the three requirements was met.

There was no communication between the towns and villages. The Sarpanches of the villages had no means of informing the district administration of the extent of damage, the number of survivors and the rescue and relief requirements. They had no transport. Whatever transport was available was used to evacuate the casualties. I am also sure that they had no idea as to whom they should contact. All that they could do was to wait for the Army, Paramilitary Forces or NGOs to reach their village and start the relief process.

One can also be reasonably sure that no one at Bhuj knew where each of the 900 odd villages were located and what was the route to them. The Army and Paramilitary Forces may have maps which would enable them to find the villages. The best an NGO could have is a road map of Gujarat. Such maps do not show most of the villages. Hence even if one wanted to send the relief material to the villages, the task was almost impossible.

Where there is no control or coordination, there has to be confusion. Confusion was very much in evidence in distribution of relief. Stocks piled up at Bhuj. There was no organization to take control and distribute. Towns and villages on the highways got some relief. Those in the interiors were forgotten for days. A relief commissioner and assistant relief commissioners made an appearance after a few days. But the organization required to enable them to function did not exist.

Government is not Responsive to the Plight of the Villages

This is also true. In our country, the urban rural divide and the rich poor divide is there for all to see. The attention of the Government is on urban development. Many villages are without drinking water. I wonder how many collectors can claim to have visited every village of his district. Even during normal times, the villages are neglected. At the times of calamities, they are often forgotten.

Government had Allowed Illegal and Unsafe Construction

This has been established beyond reasonable doubt during the earthquake. We have laws but are not willing to implement them. For a few rupees, all categories of law enforcers are willing to look the other way. The rich and the influential flout all rules with impunity, secure in the knowledge that they are above law. During my four days trip of Kutch district, I could not find a single structure that satisfies all the requirements of earthquake resistant construction as applicable for Seismic Zone V. The Government servants are not liable for their acts of omission and commission.

Conclusion

The role of the Government of Gujarat in the aftermath of the earthquake left much to be desired. Some of the expectations of the public and the media are beyond the capability of any government. But the absence of a realistic and implementable Disaster Management Plan was largely responsible for the sorry state of affairs. The failure to ensure that all construction in the district was earthquake resistant aggravated the situation.

9

LESSONS FROM THE GUJARAT AND OTHER EARTHQUAKES

Kutch region of Gujarat is located in Seismic Zone V. Adjoining Saurashtra is in Zone IV. The regions have been regularly devastated by earthquakes every 40 to 50 years. Anjar and five other villages were rebuilt after the Earthquake of 1956. Yet the buildings constructed in the area are not constructed to be earthquake resistant. All the buildings reconstructed in the aftermath of the 1956 Earthquake have collapsed. The houses, which stand intact in splendid isolation amongst the ruins, testify to the fact that the gruesome loss of life and property is not so much due to the earthquake itself but to the defective designs and poor quality of the structures that existed in the area. We never seem to learn from past mistakes. It would perhaps be appropriate to learn lessons from this tragedy and ensure that the houses, which are reconstructed in such areas, are of appropriate design and quality and the magnitude of the devastation is prevented.

Design of Structures

Shape of Buildings. In Seismic Zones III, IV or V, the shape of the buildings should be symmetrical in the horizontal plane like square, rectangular, circular, hexagonal etc. Irregular or asymmetrical shapes like 'L' or 'H' are more prone to damage during earthquakes. The buildings should also be balanced and symmetrical in the vertical plane or elevation. In simple terms, the building should be wider at the ground floor level or at least the same width as the upper floors. Buildings, which are wider at the top are more unstable.

Load Bearing Structures: All buildings can be broadly classified into two types from design point of view. These are 'Load Bearing Structures' and 'Framed Structures'.

Load bearing structures are those buildings where the roof is supported on the walls. The roofs can be of concrete, stone slabs, tiles, CGI or Asbestos Sheets or even thatched. The walls of the building can be of bricks, stones, concrete blocks, mud etc. The mortar used for bonding of the wall material can be cement mortar, lime mortar or mud mortar. Load bearing structures can be single, double or three storied. There can be beams of RCC, steel or timber in these houses but the ends of the beams rest on the walls. All load bearing structures must be provided with 'bands' as recommended in the relevant IS Codes. The other considerations related to the size and position of openings and corner reinforcements must be followed.

Framed Structures: Framed structures are those structures where the roof is supported by a frame and not by the walls. The frames can be of RCC, steel or timber. The walls fill in the frames. The walls of the building can be of bricks, stones, concrete blocks, mud etc. The mortar used for bonding of the wall material can be cement mortar, lime mortar or mud mortar. Framed structures can be single, double or multi-storied. All framed structures must be designed by qualified, experienced and competent structural engineers. In doing so, the structural engineer must certify that in carrying out the designs, the relevant IS Codes on earthquake resistant construction have been followed.

Ceiling Plaster: Ceiling plaster should be avoided in Seismic Zones III, IV or V as they tend to fall during earthquakes and injure people. If used, they must be kept thin.

Design of Load Bearing Structures

Provision of Bands: Bands are mandatory for all load bearing structure in Seismic Zones III, IV and V. There are four types of bands which may be required depending on the type of structures. **Plinth** and **lintel** bands are mandatory for all load bearing structures. **Roof** band is necessary when the roof height is more than 4 m. **Gable** bands are necessary in case of buildings with sloping roof with gable walls. Bands must be continuous throughout the building and on all walls. Otherwise they are not effective. The bands are made of 1 : 2 : 4 reinforced cement concrete. Special attention has to be given to the tying of reinforcements at the corners and junctions of walls. The band must be at the same level on all walls. The details are given in Chapter 15.

Design of Walls. While designing the walls the following points should be kept in mind:

- Height of walls should be kept to the minimum acceptable.
- Length of any wall should not be more than 5 m.
- Total width of openings in any wall should not exceed half the total length of the wall. All openings should be at least 60 cm away from corners and the gap between two openings should be at least 60 cm.
- Gable walls of barrack type buildings with sloping roof should be strengthened.
- Where stone masonry walls are proposed, 'headers' must be provided as laid down in the IS Code on stone masonry. 'Corner stones' must be used and the corners of the buildings must be properly detailed. The thickness of walls should be 30 cm. The walls must be constructed in cement mortar and properly cured.

Design of Roofs. Heavy RCC sloping roofs should be avoided. The weight of sloping roofs produce a horizontal thrust on the walls. This thrust increases during earthquakes and can cause the wall on which it is resting to collapse. Cracks along the ridge were noticed in many houses with two-way sloping roofs in houses built during rehabilitation of victims of Latur Earthquake. Stone, slate and tiles should not be used as roofing material. RCC or CGI sheets are the most suitable.

Design of RCC Framed Structures

Use of IS Codes. RCC framed structures, whether two storied or multi-storied, should be designed by competent structural engineers. It must be ensured that the relevant IS codes on construction of earthquake resistant buildings are followed.

Avoid Stilt Parking. Today, almost everyone has a car. Parking is a problem and stilt parking is popular. However, in case of stilt parking, there are no filler walls between the columns. The ground floor is thus less rigid than the upper floors and tends to give away during earthquakes. This was the case with most of the buildings at Ahmedabad. Hence stilt parking should be avoided.

Avoid Floating Columns. Floating columns are those columns which do not have a foundation and do not start at the ground floor. These are favoured by architects for prestigious buildings like hotels and commercial complexes. These can be dangerous.

Avoid Cantilevers. Cantilevers and projections are also favoured by architects. However, in Seismic Zones III, IV, or V, cantilever projections more than 1.5 m are not advisable.

Design of Walls. In framed structures, the walls are not load bearing. However, the bond between the frame and the wall is always weak. These walls crack and sometimes collapse during earthquakes. Hence the considerations applicable for design of walls for load bearing structures should be followed.

Considerations for Foundations

Construction on filled up ground. Earthquakes shake up the foundation. Where the foundation rests on filled ground, the chances of settlement during earthquakes are very high. Hence buildings, particularly large ones, should not be on filled ground. Whenever this is unavoidable, piles should be used to rest the foundations on hard strata.

Subsoil Investigations. Except for single or double-storied light construction, subsoil investigations must be carried out to establish the depth of founding strata and the safe bearing capacity.

Depth of Foundation. Except for light weight single or double storied houses, the foundations should be designed based on the data obtained from subsoil investigation. In case of low cost housing a 1.8 m deep open foundation should be adequate.

Width of Foundations. Width of foundation for low cost housing could be kept at 0.9 m.

Precautions in Black Cotton Soil. Black cotton soil expands and contracts to a large extent depending on moisture content. Hence special precautions are necessary when constructing houses on such soil. In case of non-engineered structures, the depth of foundation should go below the line of cracks. The width should be 30 cms more than what is given above. The gap between the sides of the foundation masonry and the sides of the foundation trench should be filled with sand or morum. One metre wide plinth protection should be provided to reduce chances of water percolating to the foundation.

Water Table. Where the depth of ground water is less than 3 m, construction should be avoided. If this is not possible, a specialized foundation engineer needs to be consulted.

Quality Management

Quality management during construction is most important. Best designs will fail during earthquakes if the quality of construction is bad. Quality depends on quality of material and quality of workmanship. Material must be tested on receipt at site and material not conforming to laid down specifications must be rejected. Some aspects regarding quality of workmanship are discussed in the subsequent paragraphs.

Construction of Masonry Structures

Brick Walls. The following aspects must be ensured:

- Bricks must be well soaked before use. Over night soaking is desirable.
- Bricks must be laid frog upwards.
- Standard bonds must be followed and staggering of joints must be ensured.
- Only that much mortar which can be used within half an hour should be mixed at a time. Otherwise the cement in the mortar will set and the mortar will lose its strength. Mortar joints should not be more than 12 mm thick.
- Proper bonding must be ensured at the corners and at junction with other walls.
- Line, level and plumb must be maintained.
- Masonry must be cured with water for a minimum period of 7 days.

Stone Masonry Walls. The following aspects must be ensured:

- Header or through stones must be used. If they are not available, reinforced concrete headers, 15 cm by 30 cm and same length as the width of the wall should be cast with one to two reinforcement bars and used. They should be spaced in rows, 1.2 m centre to centre in the horizontal direction. The vertical distance between two rows should be 0.9 m.
- Specially dressed corner stones must be used at the corners. The width of a corner stone should be equal to the width of the wall. Its length should be twice the width of the wall. The corner stones in adjacent courses must be at right angles to each other.
- Small or round stones should not be used. The minimum dimension of a stone should be 15 cm.
- The stones must be laid in cement mortar. Only that much mortar which can be used within half an hour should be mixed at a time.
- Proper bonding must be ensured at the corners and at junction with other walls.
- Line, level and plumb must be maintained.
- Masonry must be cured with water for a minimum period of 7 days.

Cement Concrete Blocks

These can be used in lieu of bricks in walls. The blocks should be of regular shape and tested before use. However, it was noticed during the visit to the earthquake affected area that the performance of CC block walls were inferior to brick walls or stone masonry in cement mortar. This is because the blocks do not have any frogs and hence the bond between two blocks is not strong. The walls tend to shear in the horizontal plane along the joints. Dressed sand stone has the same problem. Severe leakage problem was also noticed in most of the CC block walls in houses reconstructed after the Latur Earthquake. CC blocks or dressed stones should be the last option.

Construction of Reinforced Concrete

Quality Control of Concrete. This is most important. Without proper control concrete does not achieve its design strength and is likely to fail during earthquakes. Material used and cement must be tested before use. The following aspects need special attention.

Water Cement Ratio. This is most important. Excess water improves workability but reduces strength. Telling the water cement ratio to the mason in charge has no meaning. He needs to be told how many litres of water should be added to the mixer for each pour. He must have a can with which the quantity of water can be measured. Water cement ratio must be strictly controlled and tested by the slump test.

Silt Content in Sand. Presence of more than 8% silt in sand by volume weakens concrete. Sand should be tested and if the silt content is more, 5% extra cement should be added.

Curing. This is absolutely vital. Slabs should be cured by ponding water on top of the slab for a minimum period of 14 days. Columns should be cured by wrapping them in Hessian cloth and watering them for 14 days. Vertical props supporting the slab should not be removed before 14 days.

Use of Correct Mix Design. M 15 concrete should only be used for non-engineered structures. The design mix of 1 : 2 : 4 should be strictly followed. For designed RCC framed structures, the quality of concrete should be laid down by the designer. The mix design must be worked out and strictly followed. Six test cubes must be taken for each days work and tested for 7 days and 21 day strengths. Results must be critically examined and appropriate actions should be taken.

Hacking of Columns. This must be done as soon as the shuttering is removed to improve the bond between the columns and walls.

Correct Tying of Reinforcements. Reinforcements must be correctly tied as per the bar bending schedule provided by the designer. This should preferably checked and certified by a representative of the structural engineering consultant a day before actual concreting. In addition to checking the number, size and shape of the bars, the following aspects must be checked:

- Overlaps at joints of reinforcements must 52 times the diameter of the bar. It must also be ensured that not more than one-third of the reinforcement bars in any member terminate at the same point.

- 'Ghora' or support must be provided to cranked up bars to ensure that they are not pressed down by the labourers when pouring the concrete.
- Cover blocks of the right size must be pre-cast and placed under the reinforcement bars.
- The shuttering must be properly erected and the props must be of the correct size and spacing and firmly wedged on hard ground or flats. Gaps in the shuttering should be closed with packing tape to ensure that the slurry does not escape. The top of the slab must be clearly marked on vertical shuttering at the edge of the slab or beam. This is to ensure that the thickness of the slabs or depth of beams is as per design.

Use of Vibrators. Skid vibrators must be used for compacting concrete laid in slabs and pin vibrators must be used for compacting concrete in columns. Otherwise there would be voids in the concrete and the structure would be weak.

Temperature Restrictions. Concrete should not be laid when the temperature is more than 40 degrees centigrade.

Construction of Timber Framed Structures

Timber framed construction are common in some rural areas. These are safe and acceptable. However, some precautions must be taken. These are discussed in the subsequent paragraphs.

Bracing of Joints. All junctions in timber frames should be braced as shown in the diagram, both in horizontal and vertical directions. The bracing can be of timber or MS flats or angles of minimum size of 50 mm or 2". They should be screwed or bolted to the timber members. The details are given in Chapter 15.

Securing of Timber Posts. The base of the column should be rigidly fixed by one of the two methods shown in Chapter 15.

Conclusion

Appropriate design and good quality construction are the best guarantee against collapse of buildings during earthquakes. One without the other is useless. A chain is as strong as the weakest link. Excellent quality of construction will not prevent a building from collapsing if the design is not earthquake resistant and appropriate to the Seismic Zone and the foundation strata. Appropriate designs will fail if the quality of construction is poor.

To achieve appropriate designs, we need to curb architectural flair and emphasis on appearance and concentrate on designs which are earthquake resistant. In our country architects rule the roost and appearance is of primary importance. Structural engineers take the back seat. Pragmatists are inconvenient and ignored. Designs must take into account ground realities and construction problems. It is difficult and costly to maintain quality of concrete in mass housing in hot and arid areas of Kutch or Marathwada. Foreign consultants advised against use of CGI sheet in the relocated villages in Latur because of the heat. They are not aware that 500 million Indians in rural India have lived for centuries without fans in temperatures of over 40 degree centigrade and most of them still do. Heat for them is a very minor consideration. Safety is their main concern.

To achieve quality of construction we require knowledge, systems and commitment. A school recently constructed by a religious trust collapsed at Ahmedabad during the earthquake killing 70 children. I am sure there was no lack of commitment to quality on part of the trust. But lack of knowledge was fatal. It must also be appreciated that the requirement of knowledge at different levels is different. A labourer mixing concrete does not need to know the water cement ratio. He should know how many litres of water he needs to pour into the mixer and must have the wherewithal to measure the water. A labourer tying reinforcement does not need to know that the overlap should be 52 time the diameter of the bar. He needs to know the length of the overlap. Most of the masons and artisans who work at the worksites are illiterate and need to be explained what they need to do in very simple and clear terms.

The systems include testing of materials and workmanship and maintaining proper records of the same are essential. Quality assurance also requires preparation of checklist for monitoring activities in the preparatory stage, construction stage and post-construction stage.

Commitment implies the will to enforce the systems in the face of all odds, pressures and inducements. Without commitment to quality from senior managers to the junior most site person, quality can never be achieved.

If we are to avoid repetition of the havoc caused by the Gujarat Earthquake of 26th January 2001, we must ensure that whatever is rebuilt is correctly designed, properly constructed and earthquake resistant.

PART III

What We Can Do

10

PRELIMINARY SIGNS AND PREPARATIONS

It is true that it is not possible to scientifically predict earthquakes. But sometimes there are indications. These include foreshocks and astrological predictions, which can be taken seriously. Small tremors are often felt before a big one. So if your risk analysis indicates that your overall risk is moderate and above, it would be prudent to make some preparations if small tremors are felt. If there is an astrological prediction of an earthquake, you can take some leave and visit your relative or friends staying elsewhere or simply go on a holiday.

But if you cannot get away, is there anything we can do? I believe there is. There are some precautions, which we can take and these do not always cost money. These are called contingency plans and could include the following:

- Identify the safest place within the house.
- Stock up.
- Emergency kit.
- Emergency repairs to your house.

Safest Place in the House

When earthquakes come, it is said that the best thing to do is to get out of the house and out into the street or open spaces. This is not always possible. Earthquakes last only for about a minute or maximum two minutes. If it is not possible to get out within a minute, there is no point in trying. One should not panic and jump out of first or higher stories and court certain death as some did in Ahmedabad and Surat. Also do not use lifts as they are likely to get stuck during the earthquake.

Getting out of the house is also no use unless the streets are wide and the buildings lining the streets will not fall and kill those who come out. Over 400 children and 60 teachers parading along the narrow streets of Anjar were killed when buildings collapsed and fell on them. So outside the house is not always safe. We should act as the circumstances dictate. If we are staying in the ground or first floor and there are wide streets or open spaces just outside the house, the best thing to do is to rush out of the house. We should move out to be sure that we can reach the open spaces within 60 seconds. If that is not possible, it is better to stay in the house.

If the plan is to stay within the house in case of earthquakes, every member of the house must know which is the safest place in the house and move there as soon as the tremors start. One might also sleep there at night if frequent light tremors are being experienced. It is said that places under beams and door frames are safest and are the places to go to. Unfortunately, in modern buildings, the walls are under the beams. So there are very few visible beams if any. The space under a door frame is too small to accommodate people. Getting under beds and tables is all right. But today's furniture are not as strong as they used to be. So what are we supposed to do? My suggestions are as under:

Avoid the following:

- Avoid large rooms. Larger the room, greater is the chance of collapse of the slab.
- Avoid projecting balconies or rooms where the projection is more than 1.5 m.
- Avoid rooms, which are on or adjacent to floating columns or columns which do not have foundations but start on the first or upper floors.
- Avoid rooms on the outer face of the building. These have the least rigidity and hence most susceptible to damage.

Select the following as the safest place in the house:

- The narrowest space between two beams. To identify this space you may have to look at the structural plan or consult a civil engineer. If your building has stilt parking, an examination of the column and beam arrangement in the ground floor will give you an idea. The likely areas could be passages, kitchen, study or any other small room.

- It will be desirable to be close to the lift or stair well because these are usually stable structures.

Stocking Up

If we survive the collapse and are trapped in a pocket, it may be days before we can be rescued. To survive the period we need some food and water. It would be a good idea to stock some emergency rations and water in a box or trunk in the safe area of the house. Dates and chocolates are the best emergency rations as they contain the maximum calories per gram and do not spoil easily. Gur or jaggary is also a viable option. The quantity to be stocked depends on the number of people in the house.

Emergency Kit

In addition to the food and water a few other things could be very useful. These are torch and torch cells, candles and matchsticks, a first aid box and a crowbar. The crowbar is most useful for breaking holes in the walls and forcing jammed doors, windows and grills in windows.

Emergency Repairs

If you happen to stay in a house which has been declared unsafe by the municipality or in a house that shows signs of distress like major cracks, find some other accommodation if you can afford it. If you cannot, you should consider getting some emergency repairs to the house. Assessment of condition of buildings and strategy for repairs have been discussed at Part 3, Chapter 13 titled 'Damage Assessment of Partially Damaged Buildings after Earthquake' and Chapter 14 'Repair of Partially Damaged Buildings'.

Conclusion

Many may consider predicting or preparing for an earthquake to be alarmist and silly. But the fact is training and preparations have saved many lives during emergencies and disasters. The cautious and the prudent may like to remain prepared for any eventuality.

It must be appreciated that panic and preparations are two different things. When we panic, we act without reasoning and often do what we should not do like jumping off multi-storied buildings to certain death or grievous injury. Training and preparations are reasoned and deliberate. They are the best methods of disaster mitigation.

11

RISK ANALYSIS AND PREPARATIONS

What Can We Do

Man is almost powerless against the fury of nature. We cannot always change the place where we live and work. We cannot predict earthquakes and run away from it. We should not live in fear. The earthquake may not kill us. But if we live in constant fear of earthquakes, the mental stress that is generated will certainly kill us in the long run. So what are we supposed to do? We are supposed to take measures to mitigate the possible effects of the earthquake.

Mitigation measures are those measures, which are taken to reduce the effect of disasters. Some are required to be taken by the Government like construction of a flood control embankment or a cyclone shelter. Others are required to be taken by individuals. To give you an example, my brother-in-law stays in a flood prone area of Assam. Having once spent three days with his family on the sloping tin roof of his house and lost most of his important documents, he decided to take mitigation measures. He built a small ground plus one annexe with a small kitchen, bathroom and water storage facility. All his important papers and food stocks are stored on the first floor. So if there is a flood, all he has to do is to move to the annexe and hold out till the water level comes down. Mitigation measures cost money. But even the poor, if they understand the risk, can do something about it. For example, many of the victims of the earthquake at Latur do not sleep in the concrete houses built by the Government for them. They have constructed out houses with CGI sheet roofs and sleep in them.

Since mitigation measures cost money, it is important that we analyze the risk and decide what measures we need to take.

Risk Analysis

The first step in deciding the mitigation measures to be adopted is to carry out a risk analysis. The risk analysis must try to quantify the risk to life and injury, risk to property and risk to documents. All these risks are actually related to the ability of your residence and workplace to survive an earthquake. If the risk of your residence or office building suffering severe damage during earthquake is low, the chances of loss of life, limb and property will be low. To assess the risk we should consider the following factors:

- Earthquake zone in which we are located.
- Whether we are located astride faults or not.
- Subsoil conditions of the place where we have our residence or work place.
- Type of house in which we live or work.

To enable us to try to quantify the risk, let us quantify each risk factor into five degrees or magnitude of risk.

Seismic Zone Risk. The risk from earthquake depends on the earthquake zone in which we are located. India is already divided into five earthquake zones. To assess the magnitude of risk from this factor, we should look at the Seismic Zoning Map given at the end of Chapter 1 and identify the seismic zone in which we live. The risk related to seismic zoning is as under:

- If we are residing in Seismic Zone I, the earthquake, which may occur is likely to be of intensity less than 5 on the Richter scale. Earthquakes of this intensity do not cause major damage to buildings. Occasionally, minor cracks may be seen. Hence the degree of risk due to seismic zone is graded as 1.
- If we are residing in Seismic Zone II, the earthquake, which may occur is likely to be between 5 to 6 on the Richter scale. An earthquake of this magnitude is likely to cause light damage in the form of plaster cracks, separation cracks and stress cracks may appear in walls. Structures are unlikely to collapse. The risk in this seismic zone is likely to be graded as 2.

- If we are residing in Seismic Zone III, the earthquake which may occur is likely to be up to 6.5 on the Richter scale. An earthquake of 6.5 magnitude will cause moderate damage to buildings in the form of major cracks and partial collapse. Total collapse of buildings close to the epicentre cannot be ruled out. The risk in this seismic zone is graded as 3.
- If we are residing in Seismic Zone IV, the earthquake, which may occur is likely to be 6.5 to 7 on the Richter scale. An earthquake of this magnitude will cause heavy damage to buildings in the form of partial collapse or total collapse unless the buildings are earthquake resistant. Total collapse of buildings close to the epicentre cannot be ruled out. The risk in this seismic zone is graded as 4.
- If we are residing in Seismic Zone V, the earthquake, which may occur is likely to be more than 7 on the Richter scale. An earthquake of this magnitude will cause heavy damage to buildings unless they are earthquake resistant. Total collapse of buildings close to the epicentre cannot be ruled out. The risk in this seismic zone is graded as 5.

Location or Fault Risk. The next factor to be considered is location with respect to faults. The map showing major faults in India is given in Chapter 2. We should refer to this map and decide whether we are near these faults or not. Unless the fault zone is very wide (200 kms in Kutch) we can measure the distance from any visible landmark like river or line of cliffs. The risk assessment of this factor is as under:

- If we are residing in the vicinity (within 20 kms) of a major active fault (marked on the map) in Seismic Zone V or IV, the risk factor is 5.
- If we are residing in the vicinity (within 20 kms) of a major active fault in Seismic Zone III or II, the risk factor is 4.
- If we are residing in the vicinity (within 20 kms) of a minor fault (river or line of cliffs not marked on the map) in Seismic Zone V or IV, the risk factor is 3.
- If we are residing in the vicinity (within 20 kms) of a minor fault (river or line of cliffs not marked on the map) in Seismic Zone III or II, the risk factor is 2.
- If we are not residing in any of the above areas, the risk factor is 1.

Subsoil Condition Risk. The next factor to be considered is the subsoil conditions. It has been seen that houses with foundations on hard strata suffer less damage during earthquakes compared to those located on alluvium or soil. To assess the subsoil conditions, we need to know the depth at which hard rocks are available. This can be assessed by looking at open wells in the vicinity, from those who bore tube-wells and from Government Departments like Public Works Department or Ground Water Department. Your builder will also have the data if subsoil investigation has been done for designing the foundation. The risk based on this factor should be assessed as under:

- If the depth of rocky strata is less than 0.9 m below ground level, the risk factor is 1.
- If the depth of rocky strata is more than 0.9 m but less than 1.8 m, the risk factor is 2.
- If the depth of rocky strata is more than 1.8 m but less than 5.0 m, the risk factor is 3.
- If the depth of rocky strata is more than 5.0 m but less than 10.0 m, the risk factor is 4.
- If the depth of rocky strata is more than 10 m, the risk factor is 5.

House Related Risk. The final factor to be considered is the type and condition of your house. This is the most difficult to assess. Over 80% of houses in our country are non-engineered (not designed by a qualified and competent engineer keeping the IS Code for Design of Earthquake Resistant Buildings in mind). Even when designed by structural engineers, one is never sure about the competence of the engineer or soundness of the design. Getting a design checked by another structural engineering consultant is a good practice but rarely done. Even if the design is all right, it is difficult to ascertain if the quality of construction is good. Once plastered and painted, the actual quality of construction is difficult to assess. Cracks are the main indication of the quality of construction. It requires a trained and experienced engineer to assess the quality of construction. To complete the risk analysis, we will have to lay down a procedure of evaluating the risk from this factor. The suggested method is as under:

- If your house is single storied, symmetrical in shape (square, rectangular, circular, hexagonal etc), with walls constructed in cement mortar and a light roof (CGI Sheet or asbestous sheet or reinforced cement concrete but not of stone slabs or tiles or thick layer of mud), and no room is more than 5 m long and the building has no major visible cracks, the risk factor is 1. All buildings designed as per IS Codes on Earthquake Resistant Construction and constructed with adequate quality control can also be graded as risk factor 1.
- If your house is single storied, symmetrical in shape (square, rectangular, circular, hexagonal etc), with walls constructed in cement mortar and a heavy roof (of stone slabs or tiles or thick layer of mud), and no room is more than 5 m long and the building has no major visible cracks, the risk factor is 2. Double storied houses with reinforce concrete roofs or light roof can also be graded as risk factor 2.
- Houses of irregular shape like 'L', 'H' etc. up to two stories and houses of symmetrical shape, more than two stories but not more than 4 stories should be graded as risk factor 3.
- Houses of irregular shape more than two stories but not more than four stories should be graded as risk factor 4. Multistoried buildings of symmetrical shape, without floating columns (columns which do not have foundations in the ground) and without stilt parking should also be graded as risk factor 4.
- Multi-storied buildings of irregular shape with stilt parking, large cantilevers or floating columns and all buildings with stone walls in mud mortar should be treated as risk factor 5.

Overall Risk Assessment. To arrive at the overall risk factor, we multiply the four risk factors for any given building or situation. The overall risk assessment is suggested as under:

- If the overall score is less than 15, the risk is negligible.
- If the overall score is between 16 and 65, the risk is light.
- If the overall score is between 66 and 125, the risk is moderate.
- If the overall score is between 126 and 250, the risk is high.
- If the score is above 250, the risk is grave.

The above formula is empirical and based on the damage observed in various buildings in the earthquakes at Latur and Gujarat. It has been evolved by the author. Based on the above considerations, let us now try to analyze a few examples and see how the risk analysis works.

Example 1. A single storied, symmetrical building with stone masonry in cement mortar and RCC roof located at Bhuj with hard rocks 4 m below ground level. The Seismic Zone factor is 5. Location factor is 5. Subsoil factor is 3. The house factor is 1. The risk score is 5 × 5 × 3 × 1 = 75 or moderate damage can be expected. If the same house was located on soil where the depth of rocks was less than 0.9 m, the score would be 5 × 5 × 1 × 1 = 25 or light damage can be expected. If the same building was at Jaipur, the score would be 3 × 1 × 3 × 1 = 9. Hence negligible damage can be expected.

Example 2. A four storied, symmetrical building in brick and cement mortar with RCC slabs and roof located at Delhi in Trans-Yamuna area with hard rock more than 10 m below ground level. The Seismic Zone factor is 4. The location factor being close to the Moradabad fault is 4. The Subsoil factor is 5. The house factor is 2. The risk score is 4 × 4 × 5 × 2 = 160. Hence the risk is high and extensive damage can be expected. If the same building was designed to be an earthquake resistant structure, the house factor would be 1. The risk score would be 80. Hence moderate damage can be expected. If a similar building was located on the Ridge Area with rocks available at less than 0.9 m below ground level, the subsoil risk factor would be 1. The risk score would be 4 × 4 × 1 × 2 = 32. Hence light damage can be expected. If same building was at Pune, the Seismic Zone factor would be 3, the location factor would be 2, the subsoil and house factor would remain the same. Hence the score would be 3 × 2 × 5 × 2 = 60. Hence the risk would be light.

Example 3. A multi-storied building not designed as per IS Codes on Earthquake Resistant Construction, asymmetrical, with parking on stilts located at Delhi in Trans-Yamuna area with hard rock more than 10 m below ground level. The Seismic Zone factor is 4. The location factor is 4. The sub-soil factor is 5. The building factor is 5. The risk score is 4 × 4 × 5 × 5 = 400. The risk is grave and total collapse can be expected. If the same building was designed as per the IS Codes and constructed with due care, the house factor would be 1 and the risk score would reduce to 80. The risk would now be moderate and only moderate damage can be expected. If the same building, properly designed was located in Seismic Zone III, the risk score would be 60 and only light damage can be expected. If a properly designed multistoried building located in Seismic Zone III is constructed at a location where hard rock is available less than 5 m below ground level, the risk score would be 3 × 4 × 3 × 1 = 36. In this case, the risk is light and only light damage can be expected.

It may kindly be noted that the risk and likely damage will be one grade more if the house is close to the epicentre. However, as the location of the epicentre cannot be predicted, this risk cannot be quantified.

Risk Profile

The way we react to a risk depends on our risk profile. The risk profile differs from person to person. We can divide the risk profiles into five categories as under:

- **Cautious.** This category does not like taking any risk whatsoever. They are afraid of decisions they have to take and deliberate for a long time before taking decisions. They like to over insure wherever possible.
- **Prudent.** This category understands that there is no such thing as a 100% guarantee in life and 95% assurance is the best one can hope for. They try to understand the risk factors and try to reduce the risk to the minimum within the existing circumstances. The prudent take insurances and reasonable precautions in all that they do.
- **Low Risk Takers.** This category does take chances but they analyze the risks involved and go ahead if the odds appear to be in their favour or the loss, in case one loses, is within acceptable limits. This category may take chances where the stakes are minor but do not take chances if the stakes are very high.

- **High Risk Takers.** This category assesses the risks involved but play for high stakes. They generally win some and lose a few. They are optimists and generally believe nothing will go wrong. When things do go wrong, they are usually in big trouble.
- **Gamblers.** This category believes that they are born lucky. They are compulsive risk takers and are unlikely to go in for any mitigation measures. They firmly believe that things will not go wrong and if things do go wrong, they will be able to wriggle their way out. When things do go wrong, they are usually sunk.

In this book we assume the readers are prudent or low risk takers.

Preparatory or Mitigation Measures

If, after carrying out the risk analysis, we find that our risk category is moderate or above, it would be prudent to take some preparatory or mitigation measures. Some of the things which can be done without too much expenditure are given below:

- **Insurance.** The house and household goods should be insured against earthquake damage. It is assumed that lives of wage earners are already insured. It would also be wise to have 'Mediclaim' or equivalent insurance for all family members to cover cost of treatment in case of injuries.
- **Documents.** We all have a large number of documents which are essential. These include voter's identity cards, birth certificates, marriage certificates, ration cards, fixed deposit receipts, pass books, pension papers, passports, insurance policies etc. Loss of these documents can be very inconvenient. To guard against it, it is desirable to make two sets of photocopies of these documents and leave them with two relatives or friends who live in different towns.
- **Financial Arrangements.** If one survives a disaster, money becomes very important. It is therefore desirable to have bank accounts in at least two cities. Having a few credit cards could also be very useful. Keeping too much of cash and jewelry at home is dangerous.

- **Strengthening of the House.** If our house risk factor is 4 or 5, our house is not safe in case of earthquakes. It is desirable to strengthen such houses. The method of doing so is explained later in the book.

Conclusion

Human nature is strange. Money, enjoyment and entertainment have become the focus of our lives. We can spend Rs. 1,000 on a dinner or Rs. 10 lakhs on a marriage party. But when it comes to taking insurances, paying premiums etc., we think it is a waste of money. Our memories are short. The Gujarat Earthquake has already been forgotten by the media. Most of us will forget it in a few more days. But we owe it to ourselves and our families to understand the risks and take suitable mitigation measures.

12

CONSTRUCTION AND ACQUISTION OF PROPERTIES IN EARTHQUAKE PRONE ZONES

I have built a house at Alwar in Rajasthan and we were quite proud of it. One day, after the Gujarat Earthquake, my wife asked me if our house was earthquake resistant. I had to admit that it was not. I, a civil engineer by profession, built my own house in 1995 and did not make it earthquake resistant. I was not saving money. The cost of making it earthquake resistant would not have been more than Rs. 50,000, which I could afford. The fact is that I did not know that there is a thing called 'Earthquake Resistant Construction' and that there are Indian Standard's Codes for 'Earthquake Resistant Construction'. Unfortunately, there are millions like me who do not know anything about 'Earthquake Resistant Construction'. More unfortunately, even practicing structural engineers are not aware of 'Earthquake Resistant Construction'.

I learnt about earthquake resistant construction when I worked as a consultant on the Latur Earthquake Rehabilitation project. It is necessary that all persons owning houses or planning to construct houses should be aware of 'Earthquake Resistant Construction'.

Types of Construction

If we are planning to construct our own house, we have a wide option in selecting the type of house we want to build. The choice would naturally depend on availability of funds, availability and cost of land, our life style and safety considerations. The options cover the following choices:

- **Type of Structures.** The type of structures can be broadly divided into two types, load bearing structures and framed structures. Load bearing structures are those structures where the walls bear the load of the roof or the upper floors. These structures may have beams or trusses but they do not have any columns. In framed structures, a frame is first constructed with columns and beams or trusses. The frame is then filled with the walls. The walls do not carry the weight of the roof or the upper floors. The walls however improve the rigidity of the frame. The frames can be of timber, steel or reinforced cement concrete (RCC). In theory, framed structures are more earthquake resistant than load bearing structures. However, it has to be ensured that the frames are complete, properly designed and in case of timber frames, that the joints are properly braced. Load bearing structures need bands to form a frame and to make them earthquake resistant.
- **Shape of the Buildings.** The shape of the building in the horizontal plane can be symmetrical (square, rectangular, circular, hexagonal, etc.) or irregular ('L', 'H' or other irregular shapes). In the vertical plane or elevation, they can be stable (wider at ground floor and narrowing as we go up or at least same width from bottom to top) or unstable (narrower at the base and wider at the top). Regular and balanced shape is more earthquake resistant.
- **Type of Foundations.** As far as possible, the foundations should rest on hard strata. Where this is not possible due to depth of the alluvium layer, a structural engineer must be consulted. For single storied structures with light roof, 0.9 m wide and 1.8 m deep foundations in soil will be safe. In case of 'Black Cotton Soil', the depth of foundation must go below the level of cracks. The trench of foundation should be wider than the masonry by 15 cm and this space should be filled with sand to prevent direct contact between 'Black Cotton Soil' and masonry. 1 to 1.5 m wide plinth protection should be provided all around the building to prevent seepage of water into the foundation.
- **Type of Walls.** The walls of the building can be of bricks, stones or concrete blocks. Cement mortar should be used for bonding

of the wall materials. Assam type walls are also acceptable. The choice should be governed by local availability.

- **Type of Roof.** Roof should be as light as possible. They can be of CGI sheets or RCC. Mangalore tile, stone slab or slate and mud roofs should be avoided. If Mangalore tiles are to be used for appearance or reducing heat, they should be laid over CGI sheets.

Seismic Considerations

While planning the house, some seismic considerations need to be kept in mind. These often conflict with creative impulses of architects. We will have to choose between creativity and safety. Some of these considerations are given below:

- **Number of Stories:** The house should have the minimum number of stories. Houses in Seismic Zones IV and V should not have more than four stories.
- **Shape of the House:** The shape should be symmetrical as explained above.
- **Size and Location of Openings:** Total width of openings in any wall should not exceed half the total length of the wall. All openings should be at least 60 cm away from corners and the gap between two openings should be at least 60 cm.
- **Maximum Length of Walls:** Length of any wall should not be more than 5 m. Gable walls of barrack type buildings with sloping roof should be strengthened.
- **Sloping VS Flat Roofs:** Heavy RCC sloping roofs should be avoided. The weight of sloping roofs produce a horizontal thrust on the walls. This thrust increases during earthquakes and can cause the wall on which it is resting to collapse. Cracks along the ridge were noticed in many houses with two-way sloping roofs in houses built during rehabilitation of victims of Latur Earthquake. Flat RCC roofs do not generate any horizontal thrust and hence should be preferred.
- **Cantilever Projections:** Projections more than 1.5 m should be avoided.
- **Stilt Parking:** In this type of construction, the rigidity of the ground floor is less than the upper floors making the ground

floor more susceptible to collapse. Hence this type of construction should be avoided.

Recommendations

Space permitting, construction in Seismic Zone V should be single storied or at most two storied. Buildings should be rectangular in shape and cantilever projections more than 1.5 m should be avoided. Most architects are likely to recommend concrete framed structures. This is theoretically sound but can be practically most unsound. Concrete requires strict quality control and adequate curing. In the absence of this, concrete is the worst construction material. The material is heavy and a killer. It is worth noting that all the buildings that collapsed in Ahmedabad and Surat were RCC framed structures which most experts recommend. In this connection it is worth narrating one of my experiences at Latur. A beneficiary had been given a house in a relocated village. The house had RCC roof. The beneficiary had built an out-house with CGI sheet roof and used to sleep in it. When asked why, he said, "If the concrete roof collapses, we would die but if the out house collapses, we would only have minor injuries". The common reasons why concrete does not gain design strength during construction are given below:

- Use of incorrect mix. Use of less cement or more sand to improve workability can be fatal.
- Use of excess water while mixing concrete weakens concrete.
- Not making the framework water tight leads to cement slurry escaping and reducing strength of concrete.
- Not using vibrators to compact freshly laid concrete leaves voids and weakens concrete.
- Unless cured properly, concrete becomes very weak. This requires adequate supply of water, pumps and proper supervision.
- Removing framework before 14 days can weaken the structure.
- Concreting in temperatures over 40 degree centigrade weakens concrete.

I am not an international expert but I can say with confidence that prefabricated steel structures with CGI sheet roofs and any type of cladding are quick to construct and are the safest for mass housing in rural areas in Seismic Zone V or VI. Assam type wall construction should be considered. Even timber and mud houses can be made earthquake resistant and an IS Code exists for the same.

Acquisition of Properties

If we are planning to buy a house at a place which is in Seismic Zone V or VI, most of the considerations discussed above would apply. It would also be prudent to check whether the building has been designed to be earthquake resistant. This can be done by asking a structural engineering consultant to check the drawings and calculations of the structural engineer of the builder. Unfortunately, even if the designs are safe, one can never be sure that the construction has been done as per the drawings. It is also difficult to tell whether proper quality control has been exercised during construction. Reputation and past record of the builder is perhaps the only consideration. There are some simple tests which can also indicate the quality of construction which I find quite useful. These are:

- If you can break a coconut on the edge of a concrete beam or column without causing damage to the concrete, the quality of construction is acceptable.
- If you can remove the mortar from a masonry joint with your thumb-nail, the quality of construction is bad.
- If you can make deep groves in the plaster with your thumb-nail, the quality of construction is bad.
- Horizontal deep cracks in columns and diagonal or vertical cracks in beams indicate structural weakness.

It is desirable to check the design and quality of onstruction when purchasing or hiring buildings in Seismic Zones V or VI. It is also advisable to avoid the following types of buildings:

- Buildings which are not symmetrical in the horizontal and vertical plain.
- Buildings with parking in stilts.
- Buildings which are narrow at the bottom and wide at the top.
- Buildings with large cantilever balconies or rooms.

Hiring of Buildings

When planning to hire a house in Seismic Zone III, IV or V one should try to ensure that the seismic considerations given above are fulfilled.

Conclusion

Construction or purchase of a property is a lifetime investment for most of us. It is therefore necessary to be aware if the house being constructed or purchased is earthquake resistant. It is also desirable to engage a competent structural engineer to design or check the design of the house. Quality of construction is of utmost importance. Proper supervision during construction is absolutely essential.

13

POST-EARTHQUAKE DAMAGE: ASSESSMENT OF PARTIALLY DAMAGED BUILDINGS

Whenever an earthquake of magnitude 6 and above on the Richter scale takes place, some houses are likely to be damaged or destroyed. If the damage to life and property is not much, it is unlikely that the Government will take any effort to make an organized attempt to assess the damage and advise people about the method of carrying out repairs to damaged buildings. Where the damage is extensive, it will take a long time for the Government to send experts to assess the condition of buildings and to certify whether a building is habitable or not. Conditioning of buildings in villages in rural areas will take much longer if done at all. Some of the victims of the earthquake may have houses with minor damage or damages, which can be easily repaired. They may not like to wait years for Government rehabilitation and return to their houses if it is safe to do so. This desire was found to be quite strong in Bhachau and some villages around Bhuj. But the owners had no idea of how to go about repairing these buildings. It is therefore necessary for individuals and NGOs to be able to assess the condition of earthquake damaged buildings and to carry out repair and strengthening of the damaged houses in case the owners want to do so.

As discussed in the previous chapter, all buildings can be broadly classified into two types from design point of view. These are 'Load Bearing Structures' and 'Framed Structures'. The criteria for damage assessment for different types of structures are different.

Classification of Damage

As per the IS Code on Earthquake Resistant Construction, the damage to structures are to be classified into five categories as under:

Grade 1 - Slight Damage: The buildings have fine cracks in plaster. At places small pieces of plaster may have fallen.

Grade 2 - Moderate Damage: The buildings have small cracks in walls. At places large pieces of plaster may have fallen.

Grade 3 - Heavy Damage: In these buildings large and deep cracks are seen in walls.

Grade 4 - Destruction: In these buildings, there are gaps in walls. Some parts of the building may have collapsed. Separate parts of the building loose cohesion. Inner walls collapse.

Grade 5 - Total Collapse: The building has totally collapsed.

Buildings, which have suffered Grade 1 or 2 damage, are safe to live in immediately after the earthquake. Buildings, which have suffered Grade 3 damage, should be repaired before they are safe to live in. However, some portions of the building, which have suffered Grade 1 or 2 damage will be safe to live in. Buildings with Grade 4 damage may require extensive repairs or may have to be reconstructed. Buildings with Grade 5 damage will have to be reconstructed.

ASSESSMENT OF DAMAGE IN LOAD BEARING STRUCTURES

Condition of Walls

In load bearing buildings, the weight of the roof or upper floors are borne by the walls. The condition of the walls, therefore, determine whether the building is safe to live in or not. The main indications, which indicate the condition of the walls, are cracks. The cracks may be **plaster cracks** or **through cracks**. The plaster cracks are those shallow cracks, which are only in the plaster. They may be due to poor quality of plaster or shrinkage. Plaster cracks do not weaken the structure. Through cracks are cracks, which are right through the wall. They may be **static or dynamic.**

The static cracks are those cracks, which do not increase with time and are stable. Dynamic cracks on the other hand increase with time and hence unstable. They imply that the pressure or force, which caused the crack is still active. Such cracks are dangerous and the pressure needs to be relieved.

The depth of cracks can be measured by inserting a piece of stiff paper like bond paper or piece of an envelope or a thin wire. If the depth of the crack is less than 20 mm, the crack is likely to be a plaster crack and of no consequence.

To find out whether a crack is static or dynamic, paste a thin piece of glass like the slides used in a pathological laboratory or pieces of broken windowpane with araldite on the wall. If the glass cracks within 24 hours, it is a dynamic crack. Otherwise it is a static crack. Some typical cracks, which are seen in load bearing buildings, are shown on page 105 and 106.

If the inspection of the cracks indicates that only plaster cracks are present in the wall, the damage is Grade 1 and the building is safe. If small (less than 2 m long), through static cracks are seen, the damage is Grade 2 and the building is safe. If long and multiple dynamic through cracks are seen, the damage is Grade 3 and some repairs are desirable before the buildings are occupied.

If the walls are of stones in mud mortar, there is another check that needs to be carried out. This is for bulging of the wall. During the earthquake, the stone masonry walls in mud mortar may split vertically creating bulges. Such bulges indicate that the wall is unstable and may collapse due to aftershocks. Photograph on page 105 illustrates the collapse of a stonewall in mud mortar after splitting. One side of stonewall is still standing. Such damage should be treated as Grade 3. The wall must be demolished and reconstructed before the building is occupied.

Condition of Foundation

The foundation supports the weight of the walls. Earthquakes often result in settlement in the foundation. It is therefore necessary to check the soundness of the foundation before occupying the building. This can be done by inspecting the plinth (the wall from ground level to the door sill) and floor of the building. If there are through cracks in the plinth along with settlement and cracks in the floor, settlement of foundation has taken place. Such damage should be treated as Grade 3. Some repairs are necessary before they are occupied.

Condition of Roofs

The roof or upper floors of load bearing buildings may be constructed with reinforced cement concrete, stone slabs on steel joists or wooden beams, tiles, CGI or Asbestous sheets on steel, wooden or bamboo beams, trusses, rafters and purlins. Concrete slabs may have ceiling plaster. Falling roof or plaster can be a serious threat to life and limb. Hence the condition of the roof needs to be examined carefully. Heavier the roof, greater the danger during earthquakes.

In case of concrete roofs or floors, the cracks can be linear or honey comb type. Examine the roof for cracks in the same way as we have examined the walls. If there are through cracks, the damage should be treated as Grade 3. The roof needs to be strengthened before the house is occupied. If the cracks are only in the ceiling plaster, the damage is Grade 1 and there is no danger. If pieces of ceiling plaster have fallen, treat the damage as Grade 2 and there is nothing to worry. To avoid minor injuries due to falling plaster, try to remove the entire ceiling plaster.

In case the roof is of stone slabs or wooden planks supported on steel or timber beams as is seen in many old buildings, firstly examine the beams as explained below under framed structures. If the beams are sound, check the wall where the beams rest on the wall. If there is no displacement of the beam from its old position, there is no danger. If there is displacement and cracks in the walls, treat the damage as Grade 3. The joint need to be made rigid before the house is occupied. If cracks are seen in the stone slabs or timber planks except at the joint between two stones or planks, the damage should be treated as Grade 3. The damaged stone slabs or planks should be replaced before the house is occupied.

In case it is a tiled roof, check the tiles, rafters or purlins for damage or displacement. If there is damage or displacement, treat the damage as Grade 3 and repair the roof before moving in. Tiles are easily displaced during earthquakes. Falling tiles can cause serious injuries. It is prudent to avoid tiled roofs in Seismic Zones IV or V. If used to keep the house cool, they should be laid over CGI sheets.

Overall Condition

The overall condition of the building should be treated as the worst of the three elements of the building i.e. walls, foundation and roof.

ASSESSMENT OF DAMAGE TO FRAMED STRUCTURES

Condition of Beams

In framed structures, beams bear the load of the slabs, be they roof slabs or floors slabs. The beams rest on the columns and interlink the columns to form the frame. The condition of the beam is therefore very important in determining the soundness of the structure. The beams can be of steel, timber or reinforced concrete. Steel beams should be examined for deformity. Unless deformed, they can be considered safe. Timber beams do not deform. Unless broken or with a prominent sag, they are safe. Concrete beams need to be examined for cracks in the same way as explained above. Plaster cracks in beams are not dangerous and the damage can be considered Grade 1. Diagonal or vertical through cracks in beams are dangerous. If present in only one or two beams, the damage can be considered as Grade 3. If such cracks are seen in most of the beams, the damage should be considered as Grade 4. If the cracks in the beam are horizontal, they are not very dangerous and could well be due to rusting of the reinforcement bars.

Condition of Columns

As already explained, framed structures can be of timber, steel or reinforced cement concrete or a combination of them. Columns can also be of masonry. Timber columns rarely fail. Unless broken or split, they are safe. Steel columns should be examined for buckling or bending. If buckling or bending is present, the damage should be treated as Grade 3 and repaired before moving in. In case of masonry columns, examine the columns for through

Damaged Stone Wall in Mud Mortar

cracks. If these are present, treat the damage as Grade 3 and repair the column before moving in.

If the columns are of reinforced concrete, check them for cracks. If there are deep horizontal, diagonal or honeycomb cracks treat the damage as Grade 3 and repair the columns before moving in. Otherwise the damage is Grade 1 or 2.

Junctions of Beams and Columns

In case of framed structures, the junctions between the beams and columns are very important in determining the rigidity and safety of the structures. It was noticed during my tour of Kutch, almost all multi-storied buildings, which collapsed, did so because of the failure of the junction of beams and columns at the first floor level. Hence all junctions must be carefully examined. The junctions can be between same material or between different material.

In a timber frame, check the junctions for damage. If any damage or displacement is seen, treat the damage as Grade 3 and strengthen the joints before moving in. In case of a steel frame, the joint can be either welded or riveted. If the joint is welded, check the weld for cracks. If there are no cracks, the structure is safe. If the weld has cracked, treat the damage as Grade 3 and re-weld the joint before moving in. If the joint is riveted, check the rivets for shear failure. If there is no failure,

Typical Cracks in Load Bearing Buildings

the junction is safe. If there is failure, treat the damage as Grade 3. The joint has to be repaired. In case the frame is of reinforced concrete, check the junctions for cracks. If there are no deep cracks, the junction is safe. If there are deep through cracks at most of the junctions, there is a serious design problem. Treat the damage as Grade 4 and call in an expert structural engineer to suggest the repair solution.

Condition of Slabs

The condition of the slab should be checked in the same way as explained above for load bearing structures.

Condition of Walls

In framed structures, the walls do not carry the load of the roof or slabs. Hence their condition is not very important. They are generally called filler walls. Even if the walls collapse, there may be injuries but the structure will not collapse. So even if the walls are badly damaged one can live in the building as long as one stays away from the damaged walls. The walls should be inspected in the same way as explained above for load bearing structures.

The most common form of cracks, which are seen in the filler walls of framed structures are **separation cracks** which occur at the junction between the wall and the frame. The bond between the material of the frame and the wall is usually weak and the tremors during an earthquake separate the two. This type of crack may also occur in buildings due to shrinkage particularly if the mortar joints are more than 12 mm thick. These cracks do not pose any problem to the structural strength of the building. But if the cracks are wide and extensive, the walls may collapse. It is better to repair these cracks.

Even though walls in framed structures do not carry the design loads of the building, they play an important role in increasing the rigidity of the frames of the building and thus reduce the deformation that takes place in the frame during an earthquake. Hence the walls get damaged but the frame remains safe. Ground floor parking on stilts makes the frames of the floor least rigid and hence more vulnerable. That is why most buildings, which collapsed at Ahmedabad, had stilt type parking. Stilt type parking is not desirable in Seismic Zone IV or V. Where the stilt parking is already there, it may be a good idea to construct some filler walls to improve the rigidity of the frame of the ground floor.

Overall Condition

The overall condition of the building should be treated as the worst of the four main elements of the building, i.e. beams, columns, junctions and roofs/slabs.

Conclusion

Where expert guidance is not available, individuals and NGOs may use the criteria given in this chapter to condition houses. Whenever in doubt, experts should be consulted.

14

REPAIRS OF PARTIALLY DAMAGED BUILDINGS

During my visit to the earthquake affected towns and villages of Kutch, I found many persons who wanted to repair their partly damaged houses but did not know how to do so. The repairs suggested in this chapter are simple repairs to Grade 1, 2 or 3 damage in buildings. These repairs are necessary to make the buildings habitable. These repairs can be got carried out by the victims themselves with the help of tradesmen like masons, carpenters etc. Grade 4 damage requires expert advice on the repair solution.

Repairs to Walls

Repair to Cracks Under Beams: Cracks in wall immediately under any type of beam can be due to inadequate strength of wall material (mud, bricks) or weak mortar or joints in masonry not properly staggered. To repair, support the beam at one end at a time in the same way as we do during casting. Remove 150 mm (6") of the cracked wall. Prepare a pre-cast cement concrete or dressed stone block of size 150 mm deep, as wide as the wall and as long as the width of the beam (minimum 300 mm or 12"). Lay the block on the wall on 1 : 6 cement mortar after thoroughly wetting the surface. Remove supports after one day. If pre-cast cement concrete blocks are

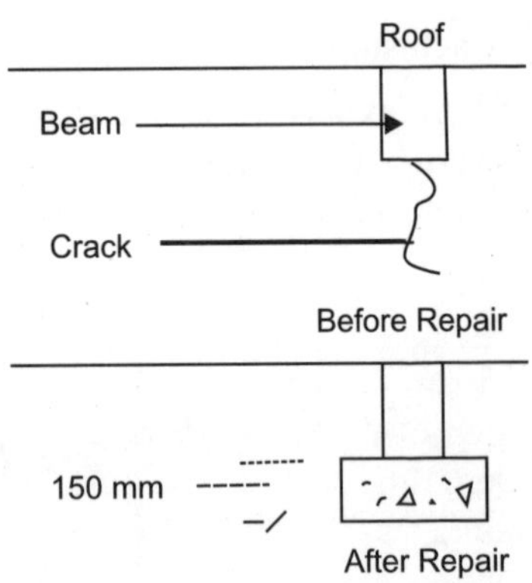

used, they must be of 1 : 2 : 4 concrete and cured for a minimum period of 14 days.

If the walls are of stone masonry in mud mortar, it is desirable to place timber or masonry columns permanently at the two ends of the beam and to put a roof band on the wall. The details of the band are given later.

Repairs to Shear Cracks: These cracks are seen mostly near openings. In brick or concrete block walls remove bricks or concrete blocks carefully at the end of the cracks and at about 60 to 90 cm along the crack. Insert new bricks or blocks in 1 : 6 cement mortar. Fill rest of the crack with crack filling material or 1 : 3 cement mortar and plaster.

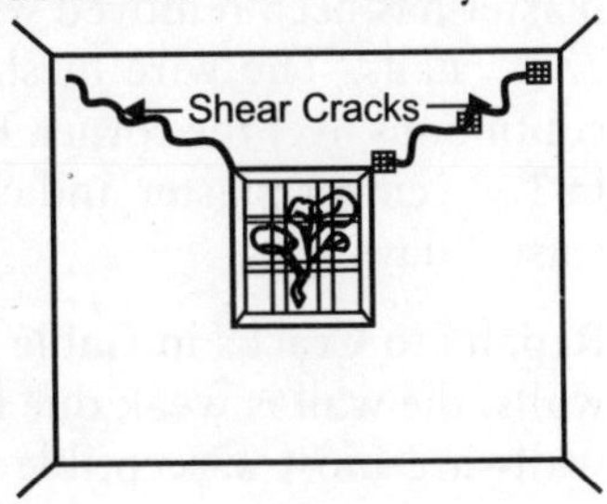

To make the repair stronger, remove the plaster from the area of the crack. Scrape the joints in the old wall to a depth of about 6 mm. Fix chicken wire mesh or 50 mm square wire mesh over the cracked area with nails. Re-plaster the exposed area with 1 : 3 cement mortar after thoroughly wetting the old wall. Cure for 7 days.

Repairs to Cracks in Corners: Cracks in corners of load bearing structures occur due to the forces created during an earthquake because the corner is weak or the bond between the two walls is not proper. In either case, if the damage is severe, it is desirable to support the roof on supports, demolish the walls at the corner for a length of 1 m and reconstruct the wall in 1 : 6 cement mortar using same material as in the existing wall. Proper bonding at corners and with old walls and curing for 7 days should be ensured. Remove the supports after 7 days. If the cracks are minor, carry out corner strengthening as explained later.

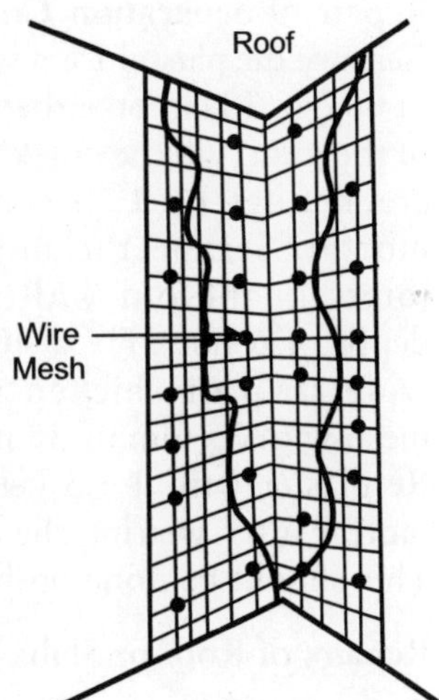

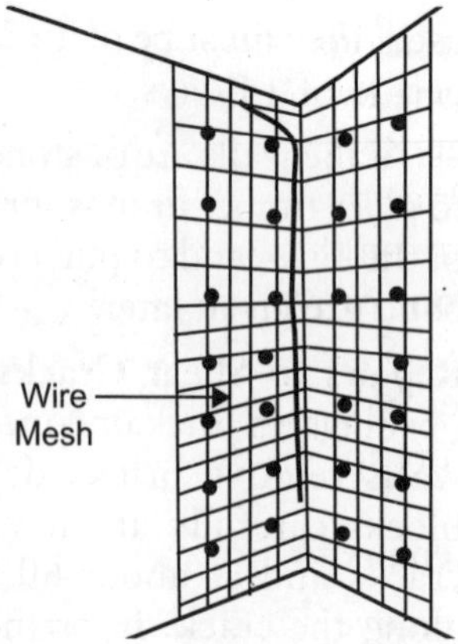

Repairs to Cracks at Junction of Walls: These occur due to inadequate bonding between the walls at the corners. Remove the plaster from roof to floor at the corner for a distance of 1m from the corner. Fix a 5 cm or 2" wire mesh on the portion where plaster has been removed with 10 cm or 4" nails. The wire mesh must be continuous over the corner. Re-plaster in 1: 3 cement plaster and cure for at least 7 days.

Repairs to Cracks in Gable Walls: In barrack type load bearing walls, the wall is weak due to its extra height and length. These walls are most susceptible to damage during earthquakes. In case of minor cracks, repair the cracks as explained above for shear cracks. Strengthen the wall at its centre by erecting a buttress on the outside. Ensure that there is proper bonding between the old wall and the buttress. This can be done by first fixing keystones or bricks into the old wall at 60 cm vertical interval or fixing 30 cm long cut pieces of steel reinforcement. 15 cms should be embedded in the old wall and 15 cm in the buttress.

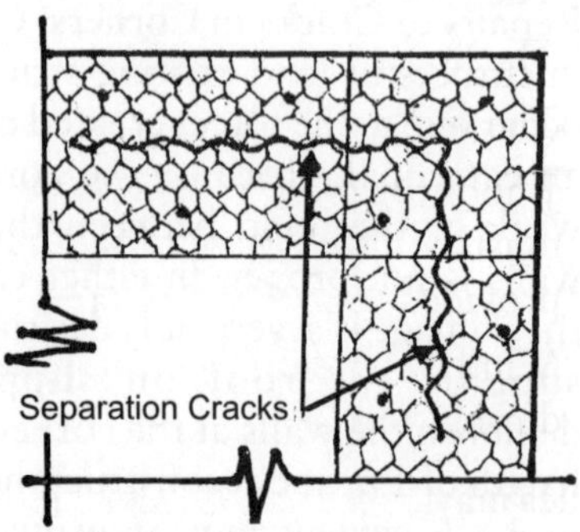

Repair of Separation Cracks: Remove the plaster for a width of 60 cm, 30 cm on either side of the joint. Fill the crack with crack filler or 1: 3 cement mortar. Scrape the mortar joints in the old wall to a depth of 6 mm. Fix a 40 cm wide strip of chicken wire mesh with 75 mm or 3" nails. Re-plaster with 1 : 3 cement mortar after wetting the old wall and cure for at least 7 days. This should be done on both sides of the wall.

Repairs of Roof or Slabs

Repairs to Concrete Roofs or Slabs: If the concrete slab shows linear or honey comb through cracks, the roof should be given additional support by inserting steel RSJs under the slab. This is done by making holes in the wall just under the slab, inserting the RSJs and closing the holes with 1 : 2 : 4 cement concrete. Once this is done, remove the brick bat coba, tiles or plaster if any on the roof and fix 50 mm or 2" wire mesh over the entire floor or roof. Cover mesh with 50 mm or 2" concrete layer of 1: 1.5 : 3 mix. In case of roof add 1 bag of water proof compound per bag of cement and give a gentle slope to the top surface. Cure for 14 days by ponding water on the roof.

Repairs to Stone Slab or Timber Plank Floors

Remove any overlay that is there over the stone slabs or timber planks. Replace the damaged stone slabs or planks. Place wiremesh and lay a 2" concrete slab as explained above.

Repairs to Tile Roofs

Tile roofs are heavy and unstable and not suitable in Seismic Zones IV or V. Remove the tiles and replace any damage purlins or rafters. Fix a CGI sheet roof on the framework with adequate J Hooks. While fixing CGI sheet roofs, the following points should be kept in mind:

- The roof should be laid from bottom to top.
- The horizontal overlap should be one and half corrugations.
- The vertical overlap should be a minimum of 30 cm.
- J hooks must be provided at the rate of two per sheet. They must be fitted at the ridges. Bituminous washers should be provided to prevent leakage.
- Wind ties should be provided to prevent damage to the sheets during storms.

Tiles may be placed on top of the CGI sheets if desired to reduce heat. Other methods of reducing heat are to cover the sheets with thatch or a thin layer of mud.

Repairs to Beams

Repair to Concrete Beams: Concrete beams are difficult to repair and require specialized skills and equipment. This should not be attempted in rural or semi urban areas unless expertise is available. If simply supported, reduce the span by building load bearing walls under it. The other option is to insert an RSJ under the beams as shown in diagram below.

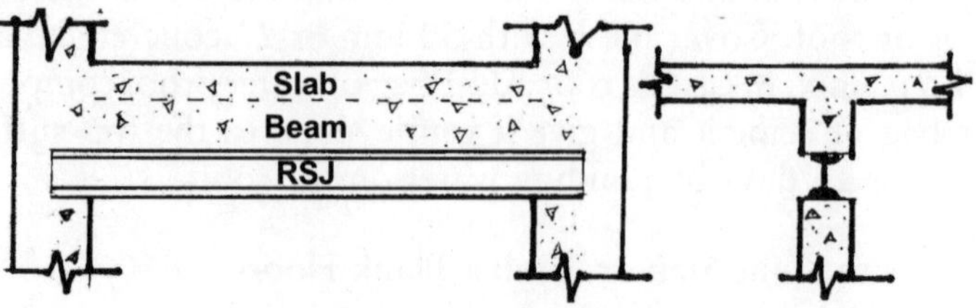

Repairs to Timber Beams: Timber beams can be strengthened by fixing mild steel plates on the sides and bottom of the beam at its centre or where there may be signs of cracks. The plate should be at least 6 mm thick, 100 mm or 4" wide and 60 cm long. The plate should have holes for fixing. Plates on the side of the beam should be fixed with 4 bolts. The plate at the bottom should be fixed with 10 screws as seen in door hinges.

Repairs to Columns

Repairs to Masonry Columns. Masonry columns cannot be effectively repaired. The beam should be supported on posts as during construction. The column should be demolished and reconstructed.

Repairs to Concrete Columns: Concrete columns are difficult to repair. This should not be attempted without expert guidance. Additional support may be placed next to the existing column to reduce the danger of collapse.

Repair to Timber Columns: Timber columns should be repaired in the same way as timber beams.

Repairs to Junctions of Framed Structures

Repairs to Junctions of Timber Frames. All junctions in timber frames should be braced as shown in the diagram, both in horizontal and vertical directions. The bracing can be of timber or MS flats or angles of minimum size of 50 mm or 2". They should be screwed or bolted to the timber members.

Repairs to Junction of Timber Beams and Masonry Columns

If there is any distress on the top of the column, the beam should be supported on stilts and the top of the column should be removed for a depth of 150 mm or 6" and replaced by the same depth of PCC. The timber beam should be fixed to the PCC base with clamps so that the joint is rigid.

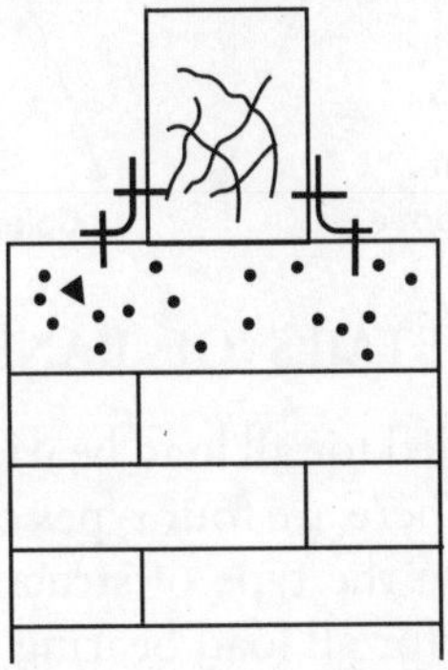

Repairs to Junction of Steel Beams and Masonry Columns

Here, the possibility of the RSJ overturning must be guarded against. Initial treatment should be the same as for timber beams. The RSJ should be embedded into the PCC as shown in the diagram.

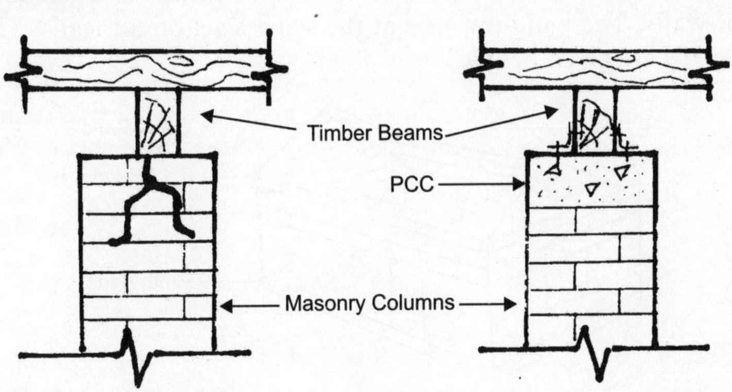

Base of Timber Columns

The base of the column should be rigidly fixed by one of the two methods shown in the diagram.

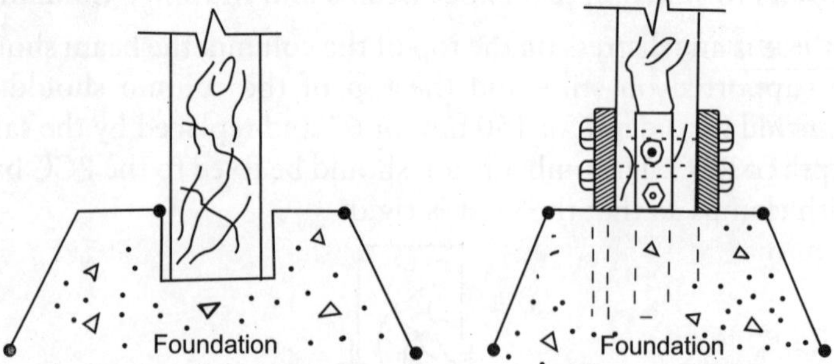

Column in Groove

Column Secured by MS Flats

DETAILS OF BANDS

Bands are recommended for all load bearing structure in Seismic Zones III, IV and V. There are four types of bands which may be required depending on the type of structures. **Plinth** and **lintel** bands are mandatory for all load bearing structures. **Roof** band is necessary when the roof height is more than 4 m. **Gable** bands are necessary in case of buildings with sloping roof with gable walls. Bands must be continuous throughout the building and on all walls. Otherwise they are not effective. The bands are made of 1 : 2 : 4 reinforced cement concrete. Special attention has to be given to the tying of reinforcements at the corners and junctions of walls. The band must be at the same level on all walls. The details are given below.

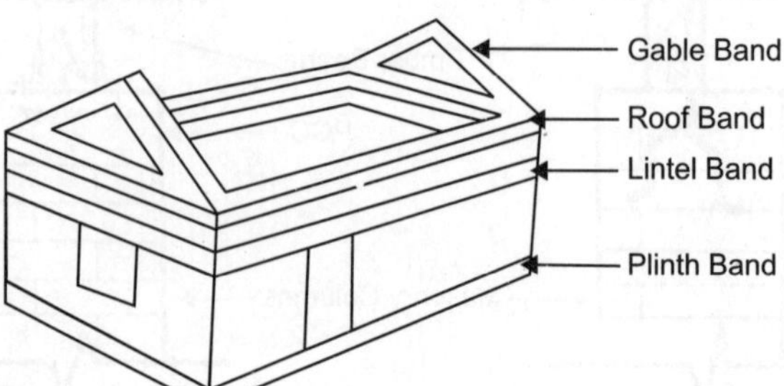

All bands cannot be provided during repairs. Only the roof band can be provided in some cases. It is advisable to provide a lintel bandage when repairing buildings. The process is explained below.

Lintel Bandage

This is provided in existing buildings where the lintel band cannot be provided. To do this, remove plaster from the outer face of the building from lintel level to roof level and deepen the masonry joints. Fix 50 mm or 2" wire mesh on the wall with nails. The wire mesh must be continuous at the corners. There should be 0.5 m overlap at junctions of wire mesh. Put 20 mm plaster in 1 : 3 cement mortar in two layer over the wire mesh. Cure plaster for a minimum of 7 days.

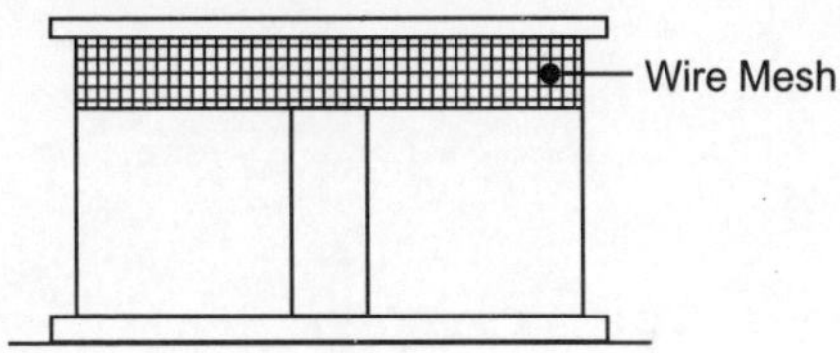

Corner Strengthening

This should also be carried out if bands are not provided in the building. To do this remove outside plaster at the corners from roof level to 150 mm below ground level for a width of 1 m. Fix 50 mm wire mesh on the exposed plaster ensuring that the mesh is continuous over the corner. Cover mesh with plaster as explained for lintel bandage.

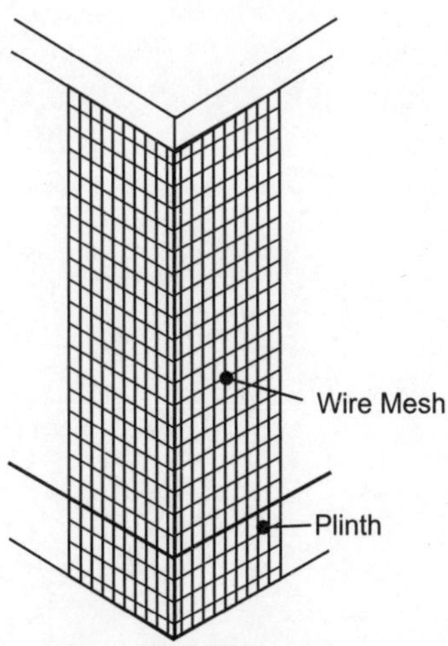

Corner Strengthening

Rules on Openings

Excessive openings (doors and windows) weaken load bearing walls and should be avoided. If the width of all openings in a wall is more than half the total length of the wall, the wall is unsafe. Some of the openings should be filled in. It should also be ensured that openings are at least 60 cm or 2' from corners and from each other.

Conclusion

Government machineries move slowly at the best of times. They may move more slowly after major disasters. Sometimes, as after the earthquake at Jabalpur, the Government may simply give some compensation and leave it to the individuals to carry out their own repairs and reconstruction. It is essential that such repairs and reconstruction be carried out as per the code of construction for earthquake resistant buildings.

15

CONSTRUCTION OF LOW COST HOUSING IN RURAL AREAS

The expertise available for planning and construction of houses in rural areas is almost non-existent. Unless the construction is done by Government agencies or NGOs, it is advisable to avoid RCC framed structures. The ability of even these organizations to maintain the necessary degree of quality control is suspect. The aspect has been discussed at length in Chapter 9. Low cost housing in rural areas should either be load bearing or timber-framed structures using locally available materials. Traditional constructions methods, modified to incorporate earthquake resistant features, is the best solution.

Load Bearing Structures

Provision of Bands. Bands are mandatory for all load bearing structure in Seismic Zones III, IV and V. There are four types of bands which may be required depending on the type of structures. **Plinth** and **lintel** bands are mandatory for all load bearing structures. **Roof** band is necessary when the roof height is more than 4 m. **Gable** bands are necessary in case of buildings with sloping roof with gable walls. The details are given below.

Bands must be continuous throughout the building and on all walls. Otherwise they are not effective. The band are made of 1 : 2 : 4 reinforced cement concrete. They should be at least 75 mm thick. The steel used should be 6 mm mild steel bars. The links should be spaced 150 mm apart. Special attention has to be given to the tying of reinforcements at the corners and junctions of walls. This should be done as per the sketched given.

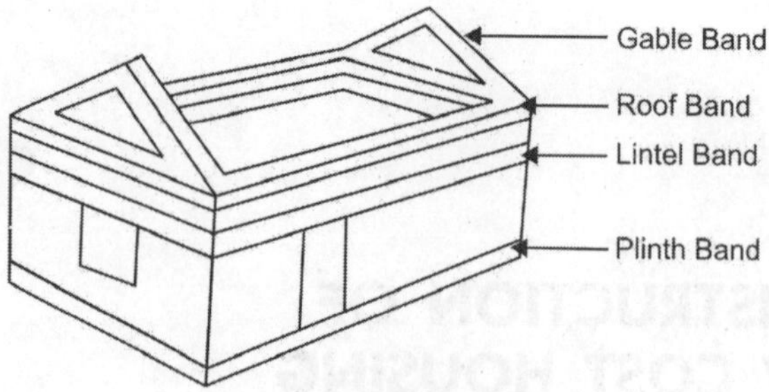

The band must be at the same level on all walls. The reinforcements of the gable band must be tied with the reinforcement of the roof band.

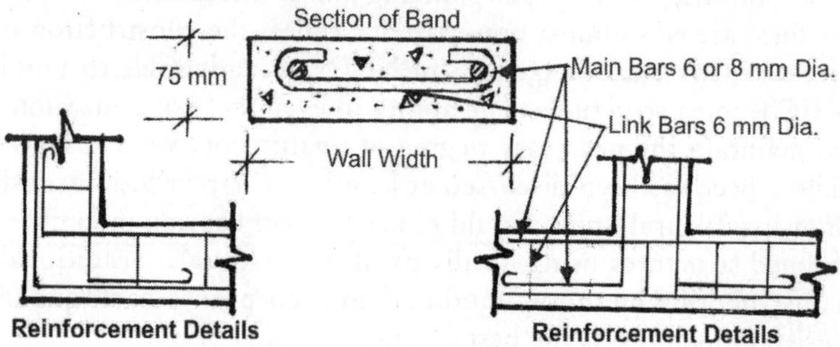

Foundation. In case of low cost housing, a 1.8 m deep open trench type foundation should be adequate. Width of foundation for low cost housing could be kept at 0.9 m. Special precautions are necessary in Black Cotton Soil. In case of non-engineered structures, the depth of foundation should go below the line of cracks visible in the excavated trenches. The width should be 30 cm more than what is given above. The gap between the sides of the foundation masonry and the sides on the foundation trench should be filled with sand or morum. One metre wide plinth protection should be provided to reduce chances of water percolating to the foundation.

Type of Walls. The walls of the building can be of bricks, stones, concrete blocks and even of mud. Composite masonry or a combination of two types of material can also be used. In case of composite masonry, the heavier material should be used in the plinth and up to window sill level. The mortar used for bonding of the wall material should be of 1: 6 cement mortar. Mud mortar should not be used.

Design of Walls. While designing the walls the following points should be kept in mind:

- Height of walls should be kept to the minimum acceptable.
- Length of any wall should not be more than 5 m.
- Total width of openings in any wall should not exceed half the total length of the wall. All openings should be at least 60 cm away from corners and the gap between two openings should be at least 60 cm.
- Gable walls of barrack type buildings with sloping roof should be strengthened.

Type of Roofs. The roofs can be of reinforced cement concrete, CGI or asbestos sheets or even thatched. Stone slabs, slate, Mangalore tiles and mud roofs supported on planks should be avoided. Roofs should be as light as possible. CGI sheets are most convenient because they are quick and easy to erect and do not require much quality control. To reduce the heat problem the sheets can be overlaid with Mangalore tile or thatch. If adequate water for curing and expertise is available, RCC flat roofs can also be constructed. However, RCC sloping roofs should be avoided. The weight of sloping roofs produce a horizontal thrust on the walls. This thrust increases during earthquakes and can cause the wall on which it is resting to collapse. Cracks along the ridge were noticed in many houses with two-way sloping roofs in houses built during rehabilitation of victims of Latur Earthquake.

Timber Framed Structures

Construction of the Frame. Timber framed construction are common in some rural areas. These are safe and acceptable. However some precautions must be taken. These are discussed in the subsequent paragraphs.

Bracing of Joints. All junctions in timber frames should be braced as shown in the diagram, both in horizontal and vertical directions. The bracing can be of timber or MS flats or angles of minimum size of 50 mm or 2". They should be screwed or bolted to the timber members.

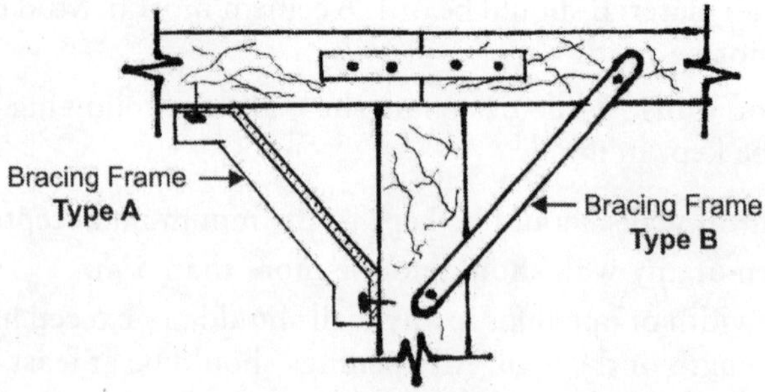

Securing of Timber Posts. The base of the column should be rigidly fixed by one of the two methods shown in the diagram.

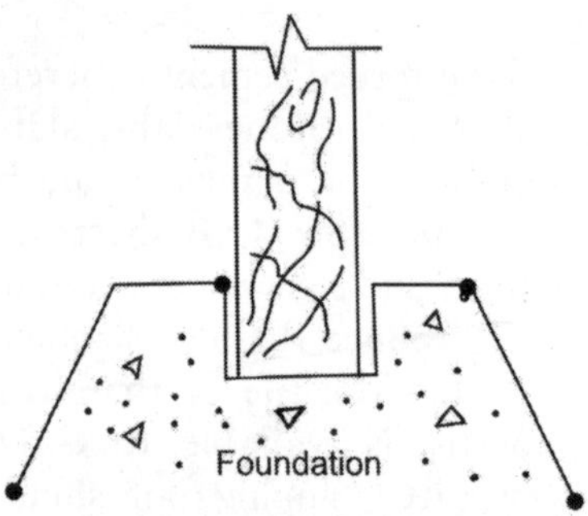

Column in Groove

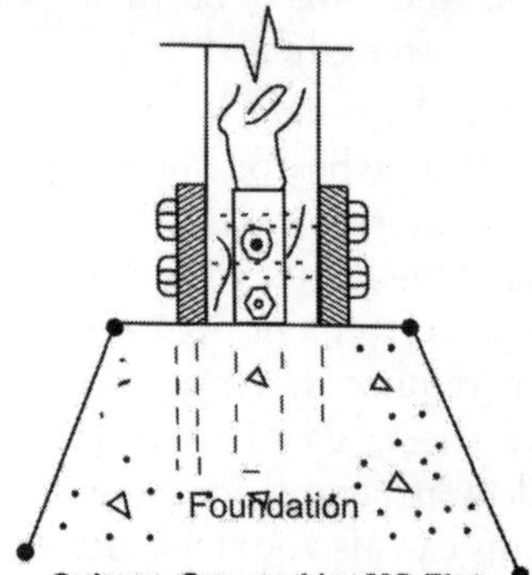

Column Secured by MS Flats

In some cases composite frames can also be used. In this type of construction, which is quite common in Satara District of Maharashtra, the posts are of concrete with a single 12 mm MS bar in the centre. The bar protrudes 200 mm outside the top end of the post and is inserted through a hole in the timber member at the top. The top end of the bar can be threaded and the joint can be made rigid with a bolt.

Type of Walls. The walls of timber or composite frame structures can be of the same type as for load bearing structures and the same considerations will apply.

Use of Bands. Bands are not necessary in framed structures. However, if the frame and the walls are independent of each other, all the three bands as applicable for load bearing structures and described earlier in the chapter must be provided.

Type of Roof. Timber framed structures should have sloping roofs of CGI sheets. Timber or steel trusses and beams should be used. Purlins and rafters can be of timber or even bamboos.

Quality Management

Quality management during construction is most important. Best designs will fail during earthquakes if the quality of construction is bad. Quality depends on quality of material and quality of workmanship. Material must be tested on receipt at site and material not conforming to laid down standards must be rejected. Some aspects regarding quality of workmanship are discussed in the subsequent paragraphs.

Brick Walls. The following aspects must be ensured:

- Bricks must be well soaked before use. Over night soaking is desirable.
- Bricks must be laid frog upwards.
- Standard bonds must be followed and staggering of joints must be ensured.
- Only that much mortar which can be used within half an hour should be mixed at a time. Otherwise the cement in the mortar will set and the mortar will loose its strength. Mortar joints should not be more than 12 mm thick.
- Proper bonding must be ensured at the corners and at junction with other walls.
- Line, level and plumb must be maintained.
- Masonry must be cured with water for a minimum period of 7 days.

Stone Masonry Walls. The following aspects must be ensured:

- Header or through stones must be used. If they are not available reinforced concrete headers, 15 cm by 15 cm and same length as the width of the wall should be cast with one to two reinforcement bars and used. They should be spaced in rows, 1.2 m centre to centre in the horizontal direction. The vertical distance between two rows should be 0.9 m.
- Small or round stones should not be used. The minimum dimension of a stone should be 15 cm.

- The stones must be laid in cement mortar. Only that much mortar which can be used within half an hour should be mixed at a time.
- Proper bonding must be ensured at the corners and at junction with other walls.
- Line, level and plumb must be maintained.
- Masonry must be cured with water for a minimum period of 7 days.

Cement Concrete Blocks. These can be used in lieu of bricks in walls. The blocks should be of regular shape and strong enough to break a coconut on it without breaking the block.

Quality Control of Concrete. This is most important. Without proper quality control concrete does not achieve its design strength and is likely to fail during earthquakes. Material used and cement must be tested if possible before use. If facilities for testing are not available, a field test can be done. Take one cup of cement and one cup of sand. Mix the two together and add a little water and make it into a ball like that of flour before making a chapatti. Flatten the ball into the shape of a chapatti, one finger thick, with your hands. Leave it to dry for six hours and then soak it in water for 24 hours. The chapatti should be quite hard. If you cannot break it with your fingers, the cement is acceptable. While making concrete, the following aspects need special attention.

- **Water Cement Ratio:** This is most important. Excess water improves workability but reduces strength. Telling the water cement ratio to the mason in charge has no meaning. He needs to be told how many litres of water should be added to the mixer for each pour. He must have a can with which the quantity of water can be measured. Water cement ratio must be strictly controlled and tested by the slump test. If engineers are not available for supervision of work, it should be made sure that the mixture does not run and there is no separation of the stone chips and fines.
- **Silt Content in Sand:** Presence of more than 8 % silt in sand by volume weakens concrete. Sand should be tested. If testing facilities or expertise for testing is not available, fill a handful

of sand into an empty bottle and cover the sand with water, shake the contents and allow the sand to settle. If a thick layer of mud becomes visible, the silt content is likely to be high. If the silt content is more, 5 % extra cement should be added.

- **Curing:** This is absolutely vital. Slabs should be cured by ponding water on top of the slab for a minimum period of 14 days.
- **Use of Correct Mix Design.** M 15 concrete should only be used for non-engineered structures. The design mix of 1:2:4 (one ghamela cement, two ghamela sand and four ghamela stone chips) should be strictly followed for roof slabs. For foundation, the mix can be 1: 5 : 10. Concrete for roof should be mixed in a concrete mixer. For foundation, hand mixing is permissible. Concrete and mortar should be mixed on an impervious surface like a tin sheet.

Correct tying of Reinforcements. Reinforcements must be correctly tied as per the bar bending schedule provided by an engineer. In addition to checking the number, size and shape of the bars, the following aspects must be checked:

- Overlaps at joints of reinforcements must be 52 times the diameter of the bar or about 40 cm. It must also be ensured that not more than one-third of the reinforcement bars in any member terminate at the same point.
- 'Ghora' or support must be provided to cranked up bars to ensure that they are not pressed down by the labourers when pouring the concrete.
- Cover blocks or flat stone about 2.5 cm, should be placed under the reinforcement bars.
- The shuttering should be properly erected and the props should be of the correct size and spacing and firmly wedged on hard ground or flats. Gaps in the shuttering should be closed with packing tape to ensure that the slurry does not escape. If packing tape is not available, banana or any large leaf or a mix of earth and cowdung can be used. The top of the slab must be clearly marked on vertical shuttering at the edge of the slab or beam. This is to ensure that the thickness of the

slabs or depth of beams is as per design. Slabs should be at least 10 cm thick.

Use of Vibrators. Skid vibrators should be used for compacting concrete laid in slabs and pin vibrators must be used for compacting concrete in columns. Otherwise there would be voids in the concrete and the structure would be weak. If vibrators are not available, the laid concrete must be properly rammed to eliminate voids.

Temperature Restrictions. Concrete should not be laid when the temperature is more than 40 degrees centigrade.

Conclusion

In rural areas, knowledge about construction practices is rarely available. Masons are generally illiterate and have experience only in traditional methods of construction. They need to be guided. This is particularly true for construction of bands and bracing of timber frames. This is also the reason why concrete roofs should be avoided in rural areas.

EPILOGUE

Knowledge and awareness about any problem amongst the people is a key requirement for controlling the problem and mitigating its effects. This has been seen in cases of diseases like AIDS and polio. Knowledge and awareness has helped in tackling the population problem of the country to some extent. Knowledge and increased awareness about disasters and disaster mitigation measures are the best way of reducing the effects of disasters.

Man is powerless in the face of the wrath of nature. Earthquakes, floods or cyclones cannot be prevented. However, their adverse effects can be considerably reduced if the people are prepared and have adopted the relevant mitigation measures.

Earthquakes do not kill by themselves. It is our lack of knowledge about the level of risk that we face and how to make our dwellings earthquake resistant, that does. It is agonizing to see that Anjar, Jowhar Nagar and some other villages of Bhuj were rebuilt after past earthquakes only to be flattened once again by the first earthquake of the 21st Century. This happened only because the new structures that were put up in place of the collapsed ones were as vulnerable as the old ones. Even Government buildings like telephone exchanges and substations in Gujarat and Indo Tibetan Border Police Complex at Mahidanda near Uttarkashi were constructed without earthquake resistant features and suffered heavy damage during the Gujarat Earthquake of January 2001 and Uttarkashi Earthquake of October 1991. Are we going to repeat the same mistakes all over again?

If ignorance is a big curse, corruption in every walk of our lives is perhaps the greatest. The greed and apathy of our politicians, the bureaucracy and government servants are all consuming. In this country milk can be produced out of urea, lubricant and detergents, edible oils can be adulterated with crippling chemicals and medicines can be adulterated. It is considered smart to evade taxes, jump traffic lights or travel in the wrong direction in a one way street and pay the ticket checker instead of buying a railway ticket. Law is likely to look the other way out of sheer apathy or for a few rupees.

It is interesting to note that most of the multi-storied buildings, which collapsed at Ahmedabad, Bhuj, Bhachau, Ghandhidham etc., during the recent earthquake, were poorly designed and poorly constructed without approvals from concerned authorities. All twelve or so hotels at Gandhidham collapsed during the earthquake, one of them before being inaugurated. Five to seven storied buildings are constructed on foundation for three storied buildings in Bhuj. Recommendations of Central Building Research Institute do not trickle down to even the Government Departments and structural engineers or consultants. No Government servant, be they ministers, bureaucrats, engineers of the municipal corporations are liable for their acts of omissions or commissions. And most of them are prepared to overlook any breach of law or non-compliance of statutory obligations for the right price.

India is a strange country. We have thousands of excellent laws but neither the political or administrative will to implement. There are over 2,00,000 RCC buildings in Ahmedabad. Only 131 of these collapsed and possibly twice that number are damaged. If intensity of the earthquake was the cause, all 2,00,000 should have collapsed. If that did not happen, the cause must lie elsewhere. The buildings were either badly designed or poorly constructed. But will those, whose negligence caused the death and injuries to hundreds of persons, be punished for their acts of omission or commission? Only time will tell.

It is sad that even after 50 years of independence, the Centre and the States do not have effective Disaster Mitigation and Management Plans and organizations. Whatever little there is, has proved to be totally ineffective. This is because these plans are prepared by bureaucrats and specialist, who have no first hand practical experience in dealing with disasters. They have no understanding of the ground realities.

Spreading knowledge is difficult. I will be lucky if a thousand buy this book and a few more read it. But rooting out corruption and lawlessness from this country is almost impossible. We have human rights groups and political patrons to protect criminals and terrorists. We have unions to protect negligent and undisciplined workers and managers. We have monthly payment systems to police and other authorities to overlook encroachments, use of polluting vehicles and running of polluting industries and a host of other irregularities. It is a government of the rich and powerful, by the rich and powerful, for the rich and powerful. Very few are interested in the common man.

The ordinary people of India have to fend for themselves. They have to learn about disaster management and take mitigation measures on their own to reduce damage and losses. NGOs can help. It is hoped that they will play a major role.

BIBLIOGRAPHY

1. *Earthquake Engineering:* Jai Krishna Volume, Indian Society of Earthquake Technology, Collection of Papers.
2. "Deathquake", *India Today*, Gujarat Special Issue, Feb 2001, Cover Story.
3. "Rebuilding Gujarat", *The Week*, Feb 18, 2001, Cover Story.
4. "Creeping Paralysis", *The Week*, Feb 18, 2001.
5. "How Safe is Mumbai", B. Krishnakumar, *The Week*, March 25, 2001.
6. "Action Time" Arun Bapat, *The Week*, March 25, 2001.
7. "The Ground Realities", Dr. Ravi Sinha, *The Week*, March 25, 2001.
8. *Report on Damage to Buildings of ITBP at Mahindanda near Uttarkashi after Earthquake 1991,* Central Public Works Department, New Delhi.
9. *Earthquake Resistant Design and Construction of Buildings—Code of Practice*, Indian Standards Institute, IS 4326: 1993.
10. *Code of Practice for Plain and Reinforced Concrete*, Indian Standards Institute, IS 456: 1978.
11. *Code of Practice for Design of Structural Timber in Buildings*, Indian Standards Institute, IS 883: 1992.
12. *Code of Practice of Construction of Stone Masonry*, Indian Standards Institute, IS 1597: Part 2 1992.
13. *Criteria for Earthquake Resistant Design of Structures*, Indian Standards Institute, IS 1893: 1984.
14. *Code of Practice for Design of Foundation in Soils*, Indian Standards Institute, IS 1904: 1986.
15. *Code of Practice for Structural use of Un-reinforced Masonry*, Indian Standards Institute, IS 1905: 1987.
16. *Code of Practice for Brickwork*, Indian Standards Institute, IS 2212: 1991.
17. *Earthquake Resistant Design and Construction*, Indian Standards Institute, SP 25: 1985.
18. *Causes and Treatment of Cracks*, Indian Standards Institute, SP 25: 1985.